Smart Mobility in Practice: Good Car, Bad Car, No Car – Is this the End of Nurturing Our Mobility DNA?

Barbara Flügge
Editor

Smart Mobility in Practice: Good Car, Bad Car, No Car – Is this the End of Nurturing Our Mobility DNA?

Springer

Editor
Barbara Flügge
digital value creators (DVC)
Bottighofen, Switzerland

ISBN 978-3-658-43170-9 ISBN 978-3-658-43171-6 (eBook)
https://doi.org/10.1007/978-3-658-43171-6

This book is a translation of the original German edition "Smart Mobility in der Praxis: Das Auto – unverzichtbar für den intermodalen Verkehr?" by Barbara Flügge, published by Springer Fachmedien Wiesbaden GmbH in 2018. The translation was done with the help of an artificial intelligence machine translation tool. A subsequent human revision was done primarily in terms of content, so that the book will read stylistically differently from a conventional translation. Springer Nature works continuously to further the development of tools for the production of books and on the related technologies to support the authors.
Translation from the null language edition: "null" by Barbara Flügge, © null null. Published by null. All Rights Reserved.

This Springer imprint is published by the registered company Springer Fachmedien Wiesbaden GmbH, part of Springer Nature.
The registered company address is: Abraham-Lincoln-Str. 46, 65189 Wiesbaden, Germany

Preface

Smart Mobility—Nurturing the Societal Mobility DNA

From my first baby steps onwards, mobility was and is in my heart. Anything in our lives, in our businesses, desires, and dreams to come true are dependent on mobility. My first book on Smart Mobility ever focuses on the most diverse currents around the topic of making mobility accessible and how intelligent mobility can thrive in cities and regions, so that it serves all of the constituents.

In the modern world, mobility drives everything—from our personal lives and dreams, to our businesses and modes of transport. In order to grow and connect as a society, we must nurture our societal mobility DNA. This means understanding modes of mobility, how we utilise mobility individually and in groups, and the needs of different cross-sections of society, including families, the elderly, and those who live in isolated country regions. Mobility has the power to change and improve our lives, and we will become even more dependent on it in the future—morphing into living spaces, business spaces, storage spaces, and travel spaces. Intelligent mobility can, and will, thrive in our cities and country regions, and holds the power to uniquely serve the needs of all constituents. This book expands on the research presented in my first book on Smart Mobility, to demonstrate how innovation in mobility can shape our lives and futures, as intelligent mobility becomes a lasting part of our societal mobility DNA.

Since *Smart Mobility—Connecting Everyone* was published, it has gained international attention. We are proud to welcome you to join our more than 400.000 and counting readers. Our readers learn about and discuss key questions such as:

- "How can we afford to travel from A to B to C?"
- "Who decides on the offers presented to us?"
- "Who provides safe transport means and sustainable infrastructures?"
- "How can struggling municipalities with tight budgets servitise their consumers by ideating new and innovative business models?"
- "What on-board resources are needed to successfully implement mobility concepts?"

- "How do you as a company in a distinct industry become a transformation mobility provider and therefore part of the smart mobility ecosystem?"

Why are these questions so crucial? An essential criterion for the success of any mobility offer is the useful, manageable, and successful project planning of this very offer for the benefit of the ecosystem.

Another essential criterion is to realise that Smart Mobility is a compound, cross-disciplinary endeavour—that has passed for quite some time the purely physical layer and invited in the digital (including cyber!) and the mental elements. That is why we call Smart Mobility a three-tier mobility: physical, digital, and mental mobility.

Our DoktorB team coordinated by Sarah (Tyrer) and DoktorB, we encountered two major findings in our trajectory of consulting projects and audits: there is a need for uniting individuals, businesses, and public administrations. Consequently, there is a need to resolve the lack of understanding in order to effectively utilise the corporate mobility assets and retail outlets to co-create a mutual offering of intermodal mobility to serve constituents locally and survive.

The fundamental question though concerns how decision-makers are able to move transformation projects onto the right track and accelerate. Transformation could take place through outsourcing the execution gap to standardised and expensive consulting concept providers. Or you as decision-makers take charge and involve your clients and community members.

Herein, you will be inviting them as participants and co-creating test users to shape the ideal offering. This could include benefits, adaptability, and even shortcomings.

Our readers reported the exercise of the Smart Mobility Ecosystem design as useful. We are running Smart Mobility Ecosystem training courses online and onsite. Take for example the New Mobility mission of the Lower Saxony constituents where we designed a service mobility concept. Other inquiries that reached us were aimed at establishing the networking strategy of a company in a series of workshops.

With regard to successful transformation, we concluded: start-ups and voluntary organisations are more open to benefit from speed coupled with innovative power, turning this into a unique selling proposition for themselves. They see an opportunity to test out and sharpen their own position in terms of the Smart Mobility ecosystem and to offer their products and services in a more targeted way. Start-ups and volunteers are more audacious than existing service providers and suppliers in known and new markets. Established companies and their strategy and research departments are certainly better skilled, better versed, and more economically viable to secure a market for themselves. Nevertheless, they need help to become unstuck already in the pilot and test phases to then proceed seamlessly. Or as we say from DoktorB's point of view: hurdles are there to be taken, not to be tolerated.

One word about autonomous driving. The discussions also revealed that public administrations and organisations are increasingly questioning their own role. Regardless of whether we are talking about providers of singular mobility, such as trains or buses, or suppliers of consumer goods, the role of regional participants is increasingly influenced by the arrival of digital assistants and developments in the field of autonomous driving. Assistants provide continuous access to new products and citizen services, for example. The accessibility of existing and new groups of buyers requires a revision of market and sales strategies.

Smart Mobility
Mobility is smart as soon as projects meet the following criteria: (1) beneficial for all involved and affected stakeholders, (2) manageable for public and private infrastructures, and (3) for the benefit of the entire ecosystem. It is not without reason that the successful implementation of intelligent mobility concepts by those that are responsible for Smart City, Smart Nation, and regional projects (Smart Regions) is one of the most important success factors to regional and urban prosperity.

In the Starting Blocks

Since the initiation of the Smart Mobility book series, we have coined the importance of ecosystems, networking, and role models. The relevance of interlinked ecosystems and smart habitats are increasingly present in articles and studies, as well as in inquiries for lectures and panel discussions. Companies such as public administrations seek cooperation to succeed in smart city and smart region projects. Sustainability, networking, and mobility were identified as key transformation triggers here. One start-up for example reached out to us to expand their business reach and serve Smart Mobility with their API in manufacturing. Smart Mobility has arrived in the application space, not just as a trend, but maturing into an orchestrator's role to help urban living spaces to survive.

Throughout this book, you will be utilising a foundation that makes intelligent mobility explainable and describable. Mobility is on everyone's lips, and the diversity of ecosystem participants opens up not only innovation potential but also questions. The review of the term Smart Mobility and intelligent mobility did not only include trends and concepts from the point of view of previous mobility providers, but rather the discourse on the fundamental change that assumes mobility supported by digitalisation.

We expanded fast. We hereby would like to thank Tim Jones, the founder of the Future Agenda. Tim, you have been such a vital sparring partner. Over time, we added more and more use cases that guided the strategic moves in cities, start-ups, the automotive industry, the manufacturing, the logistics, and other sectors.

To date, there has been no other elaboration for the German, international, and English-speaking market that has devoted itself to the topic of mobility in urban and rural living spaces in such a comprehensive way as ours.

Paying Tribute to the Diversity of Perception

Good car, bad car, no car: our committed view on the role of cars is subject to this publication. Cars have dominated for quite some time the entire evolution of technological progress, the evolution of luxury on the road, the evolution of the human being. The plethora of evolution has started with the chassis and merged into powerful electronically managed functions, e.g. braking systems, over infotainment, safety avatars towards the humanisation of the car that joins the family as a family member.

Could the car become a sustainability driver, or will it fade away by becoming the enemy of climate protection? Could the car become a new home, or will it emerge into a luxury good as OEMs like Mercedes-Benz have bet on luxury cars, abandoning their small and midsize vehicle fleets?

Since publication, the discussion about intelligent mobility and mobility for all has increased due to the question of the general availability of autonomous driving. Even the OEMs (Original Equipment Manufacturers), who are the classic automobile manufacturers, have meanwhile come to this question. Almost daily, even in relevant traditional media such as news broadcasts and print media, there are reports on the topic of the mobility of the future. A large part of the discussion is devoted to the future of the automotive industry, the classic and traditional sectors, and the combustion engine. Another discourse is on air and noise pollution and the dangers it poses to life and limb. For a long time, this discourse was overshadowed by the economic and political stability of a large number of representatives of the automotive industry. We continued to research and write about the financial feasibility of mobility to the extent that we want, need, and can afford it. In Smart Mobility Connecting Everyone, we evaluated the Mobility as a -Service concept and the go to market.

The representation of use cases and confrontation with different personae that are contributing to smart mobility have contributed to the applicability of the BIM, the Building Block for Intelligent Mobility, and their maturity assessment in mobility and transformation projects. The Procedure Model supports teams around the globe, and the overall publication has become a best seller.

This is a great endorsement for a young team and newcomer to the publishing business!

Smart Mobility in Practice. Knowing How!

Our reading community and your aspirations are as diverse as the Smart Mobility Ecosystem! In the end, successfully introduced, and above all long-term, mobility concepts should have something in common, right? For this reason, Springer Vieweg and I have decided to publish this study book, which is a practical guide. We were inspired by the case study booklets that are used in English-speaking countries for training and further education, in study projects, assessment centres, and MBA circles. This volume serves this purpose through an application-oriented, descriptive, and, above all, imitative guide.

Each accompanying use case gives the "How to know" even more space. The tools, which were developed and presented in our other books, are applied. For you, the reader, these are vividly played out through using real and fictitious case studies. Open questions and tasks are consciously discussed, even provoked: through this you will be able to determine where you stand with your own projects, what needs to be done, what might have got lost in day-to-day business discussions, and who should be your sparring partner to share and apply your knowledge that you have expanded with this book.

On behalf of the entire team, Sarah and I wish you exciting reading moments and successful implementations!

Best wishes, DoktorB

Bottighofen, Switzerland
2018 and 2024

DoktorB Barbara Flügge

Contents

List of Figures

List of Tables

About the Contributors

Horst Bishop studied computer science at the TU Vienna, where he received his diploma in 1990 and his doctorate in 1993. From 1991 to 2001 he worked at the TU Vienna, first as an assistant and then as an associate professor. Since 2001, Horst Bischof has been Professor of Computer Vision at the Institute for Machine Vision at Graz University of Technology. Currently he is also Vice Rector for Research at Graz University of Technology. Horst Bischof's research topics range from object recognition and statistical methods of machine learning to applications in the fields of surveillance, biometrics, robotics and medical image processing. In these areas, Horst Bischof has written more than 700 peer-reviewed papers. Many of the works were also awarded national and international prizes (a total of 19), including the Pattern Recognition award in 2002, the main prize of the Deutsche AG for pattern recognition DAGM in 2007 and 2012, and the British Machine Vision award 2007.

Barbara Flügge Barbara or simply DoktorB is the founder and managing director of digital value creators (DVC) Switzerland. DoktorB and her team design, evaluate, support and accelerate the transformation process of private sector organisations, Mittelstand, and the public sector.

Barbara has been helping clients increase their margins and at the same time transform into cutting-edge industries and geographies. As a thought leader, she guides the discourse on how digitisation and 360° resilience could streamline a workable transformation for urban, rural, and corporate stakeholders.

DoktorB's ABC Business Opportunity Design and OPCYBRES Cyber Resilience eLearnings are powerful transformation products. The latter, for example, helps diminish the transformational risk that companies and the public sector face. The ABC Business Opportunity Design help optimise your idea-to-market effort and speed up the launch to sell faster.

Clients are inspired by DoktorB's competences to grow and scale physically, digitally, and cognitively. As a keynote speaker and business coach, she offers her competences in the field of Digital Transformation, Smart Mobility, and Corporate Resiliency.

Clients benefit from DoktorB's industry, management, and start-up experience. She's held industry positions at companies such as Grundig and Siemens, scaled the management consultancies PriceWaterhouse and Arthur Andersen, and grown Ariba Inc.'s and SAP®'s portfolio into new markets.

Barbara holds a degree in Business Informatics and a PhD from the School of Management and Economics at Tilburg University, the Netherlands. Barbara is a bestselling author and a keynote speaker with an outstanding global reputation.

Vinod Jadhav is Product Manager for SAP® S/4HANA Cloud and Co-Innovation at SAP® SE. He is responsible for the success, quality, and delivery of S/4HANA Cloud throughout the product life cycle as defined in the portfolio, including technology trends and customer requirements to identify growth opportunities and innovations for SAP® S/4HANA Cloud. He has more than 8 years of experience in consulting business processes in various industries in the area of SAP® Financials, including payment transactions and for the financial services industry with SAP® for Insurance. Vinod Jadhav holds a Master of Science degree in Business Informatics from the Technical University of Darmstadt.

Sascha Koberstaedt is the founder and managing director of EVUM Motors GmbH. He completed his mechanical engineering studies at the Karlsruhe Institute of Technology (KIT) and the University of California, San Diego (UCSD). After practical experience in the development of electrified vehicles at BMW, he is working on his doctorate at the Chair of Vehicle Technology at the Technical University of Munich, where he will test new vehicle concepts, especially for the target region Sub-Saharan Africa.

Hans Roth started his professional career in 1983 at Becker Autoradiowerk GmbH, which was taken over by Harman International Industries in 1995. After training as a telecommunications mechanic, he studied communications engineering at the Karlsruhe University of Applied Sciences. Hans Roth has been Head of Technology Marketing for the Connected Car Division (formerly Infotainment Division) since 2012. Previously, he was responsible for Business Development and OEM Marketing.

Martin Šoltés studied aerospace engineering in Stuttgart. Since 2013, he has been researching vehicle concepts and mobility in developing countries as the chair of Automotive Engineering at the Technical University of Munich. In 2017, he founded the company EVUM Motors, which will bring a robust, simple, and modular electric vehicle to the roads worldwide.

Sarah Tyrer is a core team member of DoktorB's digital value creators (DVC) corporation and focusing on the ABC Business Opportunity Design Programme since joining.

She has a First-Class BA and Master's Degree in English Literature and Culture, both from The University of Birmingham, UK.

Christof Walter holds a degree in mechanical engineering and an MBA with a focus on lean manufacturing consulting. After graduating, he has been a long-standing manager in vehicle development and a consultant/lecturer for lean development/lean management. In 2012, he joined SAP® SE as a lean expert, where he was responsible for lean transformation, improvement, and simplification in a business unit. Since 2016, he has been working as product manager mobility, dealing with multi-/intermodal mobility concepts, shared mobility, modern drive concepts, and their effects on intelligent and smart mobility.

Mark Zannoni is research director of smart cities and transportation at IDC. As a global thought leader in urban transport, Zannoni focuses on the major and emerging transport technologies and issues affecting cities worldwide. Zannoni brings 25 years of experience in global urban transport to his research, with perspectives and insights from the public and consultative sectors, from work in Africa, Asia, Europe, the Middle East, and the United States. Its transport expertise covers a wide range of modes and topics, from traditional ones—such as aviation and rail—to innovative ones—such as autonomous vehicles and ride hailing. Mr. Zannoni is a certified urban planner and holds a master's degree in urban planning from Columbia University and a bachelor's degree in history from Boston University.

Case Study Team AIM: Marco Linpinsel and Pascal Luder

Marco and Pascal worked in the case study group of DoktorB's Smart Mobility Strategies and Management, as part of the Master Studies at ZHAW Winterthur/Wädenswil, Switzerland.

Case Study Team FAMILYIO: Gioele Panina and Nina Poglic

Gioele and Nina worked in the case study group of DoktorB's Smart Mobility Strategies and Management, as part of the Master Studies at ZHAW Winterthur/Wädenswil, Switzerland.

Terminology

aCar	Model designation for "any Car"; the vehicle which was developed within the scope of the research project of the Technical University of Munich and which is now brought to series production readiness by EVUM Motors.
ACStyria	Automotive Cluster Styria; merger of supplier and specialist companies in Styria
Alp.Lab	Austrian Light Vehicle Proving Region for Automated Driving
API	Application Programming Interface
App	Application for Internet, mobile carriers (devices, wearables, textiles)
Autonomous Driving	Autonomous driving, synonymous with driverless vehicles
BIM	Components of Intelligent Mobility
BIP	Gross domestic product
Business Network	Business network, network of stakeholders thriving for viability, feasibility and sustainability
Circular Economy	Waste-free economy
CPS	Cyber Physical Systems
Data Hub	Data hub
Decision Triangle	Individualised decision optimum from comfort, time and budget
Design Thinking	Creative method and procedure in the innovation process
Digital Concierge	Digital companion; digital assistant
Digital Mobility Front Office	Quantity of all mobility providers, digital media, consumption and interaction channels with interface to end consumer
Disruptive	Behavioural characteristic in which an existing technology, product or service is partially or completely superseded by an innovation.

E2E Mobility	End-to-end (E2E) mobility process
Ecosystems Assessment	Holistic corporate and environmental ecosystem analysis following 22 KPIs from DoktorB
eMobility	Electromobility; e-mobility
FANCI	Face and Body Analysis Natural Computer Interaction; research project within the framework of the EU funding programme Horizon 2020
HMI	Human Machine Interaction; human-machine interaction
Human Machine Interaction	Human-machine interaction
Influencer	Opinion leader
IHK	German Chamber of Industry and Commerce
IoS	Internet of Services; web of services
IoT	Internet of Things; web of things
KPC	Kids, Pets, Cargo; new user groups for autonomous driving
LRT	Light Rail Transport
LTA	Land Transport Authority Singapore
MaaS	Mobility as a service
Matchmaking	Initiation process of supply and demand
Mobility Back Office	Quantity of all providers and providers of products, rooms, services and data that ensure the deployment of a mobility service
Mobility Broker	Positioning mobility services in content and information portals and sales channels
Mobility Consumer	Mobility users
Mobility Execution	Execution of mobility services
Mobility Operations Mean	Operational means that foster mobility measures
Mobility Property	Maintenance of mobility related transhipment points and areas
Mobility as a Service	Service-driven mobility; MaaS
Moving Assets	Movable goods and equipment
MRT	Metro Rail Transport; German S-Bahn
OCCP	Open Charge Point Protocol
OCN	Occasional, Coincidentals, Nomads; new user categories for the experience space in cars
OEM	Original Equipment Manufacturer
Public Transport	Local public transport
Persona	Target audience
Physical Mobility Front Office	Quantity of all mobility providers as service and/or product and interface to customer/mobility user
Platooning	Mobile loading and unloading of trucks

Relations Mapping	Tactical and operational interconnectedness between user groups
Self-Sustaining Ecosystems	Self-supporting, viable, independent ecosystems
Service Provider	Service provider
Servitisation	Transforming physical assets into services
Smart Mobility	Intelligent mobility
Smart Tags	Intelligent markers
Smart Traffic	Traffic data analysis
Social Network	Social network
Stakeholders	Decision-makers, opinion leaders, influencers
Switchboard	Control centre
Trust Gate	Access control
USDOT	United States Department of Transportation
Use case	Scenario, definition of use scenarios are the prerequisite for field trials
User Experience	User experience
User Group	Target audience; service consumers
USP	Unique selling proposition
UX	User experience
Value Driver	Value driver

The Valuable Car

1

DoktorB Barbara Flügge

Abstract

The mobility of yesterday and today would be inconceivable without the car. The achievements that go hand in hand with the invention of the drive motor and the design of a vehicle influence our thoughts and actions, whether in everyday life, leisure time, family planning, and/or the decision to move to the countryside or to stay in the city. With innovations, the automotive industry infused our daily lives with safety, comfort and individuality. And it still dominates the economic power of a country! Should all this be called into question? How important will the vehicle be for us tomorrow? What does the car habitat look like to us? Is there a life in the car and is that life desirable?

Motivation

In this volume, we focus on the role of the car. We can think of many interesting and particularly relevant reasons for questioning the importance of the car in our planning, daily routines, smart city visions, and corporate strategies these days. These include mobility bottlenecks even in small and medium-sized cities, the diesel scandal, autonomous driving, the realisation of sustainability goals through, without or in spite of cars, the will to be surprised by the next offers from Uber, Waymo, Baidu and Tesla, or the evaluation of innovation capability and the assertiveness needed for the reorientation of internal and cross-site structures.

D. B. Flügge (✉)
digital value creators (DVC), Bottighofen, Switzerland
e-mail: barbara.fluegge@dvcconsult.com

B. Flügge (ed.), *Smart Mobility in Practice: Good Car, Bad Car, No Car – Is this the End of Nurturing Our Mobility DNA?*,
https://doi.org/10.1007/978-3-658-43171-6_1

Since its invention, the car has undergone an unprecedented transformation. Some things happen in real life, whilst some ideas and visions still take place in our heads beyond reality. The reality with which cities and regions are confronted is the task of ensuring services of general interest, i.e. the provision of mobility, and to not miss out on the proverbial connection to market events. Behind the consumers' demand for almost omnipresent mobility, there are many questions. In addition to implementation possibilities, the how, what, where, with, and who comes first considerations, test runs, presentable and scalable projects, initiatives, and initial success messages. The latter takes place using disruptive technologies and business models. Promising projects are driven by the context that the city, the mobility providers, and their users will specify for project planning.

Maintaining Mobility Is a Must

The mandate of cities and municipalities is real: they have to ensure that the mobility provision mandate is fulfilled. How much cars are needed is reflected in the considerations of cities and regions of the world that choose to pilot the implementation of mobility as a service and alternative means of transportation, as well as their examination of the advantages of electric cars and autonomous vehicles.

Focus: The Car, Our Habitat

Under this aspect of holistic mobility studies, the focus is on the car. Does a car have to be used as a means of transport and can cities manage with comprehensive, or even free, public transport without private cars? The car as a future experience space can already be seen today in the additional functions of the infotainment systems. These are offered to us as consumers and not as drivers, and depend less and less on whether we are travelling from A to B to C.

One aspect worth thinking about was covered decades ago in John Jakes' dystopia named "Firebird", first released in 1974 [1]. He describes the mobile all-encompassing space of the nomads of the twenty-first century. He refers here to a group of outlaws who are forced to live under constant locomotion, a literally mobile life. What would this life look like taking into account today's available technology and the technology that will dominate in the future? Within this idea presented by John Jakes', the concept of viable, self-sustaining ecosystems is seamlessly applicable in an ecosystem called "car". Measured against his vision, a progressive independence of life in vehicles is taking place. The terms car and vehicle are synonymous in this context with modules, moving elements, and autonomous movements.

It is the maturity of digitisation that determines if there is, or will be, a substantial or small share of urban mobility taken up by the car. Where does predatory competition between robotaxis takes place with traditional services such as buses or trams? And where do we encounter a symbiosis of all these and other mobility enablers? Who will control supply and demand in the future? Will public transport become private and shared? Who gets the bill and who the profit at the end of the day? There are no limits to the imagination.

Successful implementations, regardless of their orientation and size, count on the progress of automated driving. Then interactive and self-regulating systems in cities can actually be regulated holistically, for example by influencing pedestrian traffic, transport requirements and the replenishment control of supermarkets. Let us take a look at the stages of autonomous driving. The contents of ◉ Fig. 1.1 are based on [2].

It's Up to Us

The space that cars will offer us in the future will become living space. The efficiency of this habitat depends on the degree of automation and digitisation, and the acceptance of the ecosystem. It is up to us to acknowledge the possibilities measured against our standards.

This publication does not examine the technological depths of the combustion engine, electric mobility, or intelligent traffic light systems and laser technologies. Rather, it is a question of researching and consciously opening up a discourse on the significance of the car. OEMs such as start-ups, interest groups, and alternative suppliers are invited into the discourse.

We are interested in showing how digital advances and technological developments affect the role of the vehicle and where discontinued models are emerging. The previous role-play of manufacturers, suppliers, challengers, and vehicle owners is being drastically

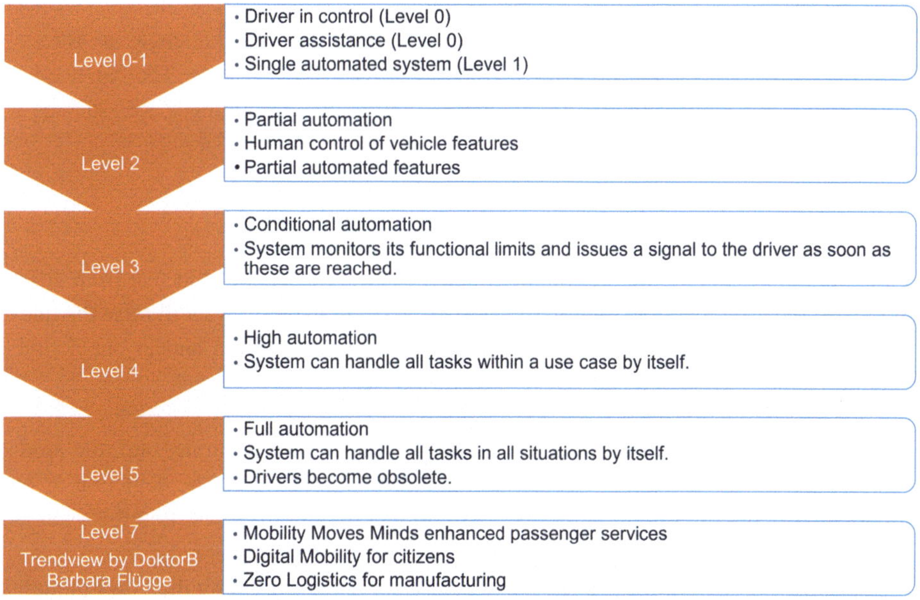

Fig. 1.1 Progressive automation as a foundation for new living spaces

challenged by a new player: the car. The proverbial independence of a thing in the sense of the Internet of Things (IoT) can already be found in offers such as the intelligent refrigerator, the electric hotel butler, or the parcel robot, which independently and autonomously delivers the order to our door. Another example is the PaketButler [2]. This ensures the secure storage and opening of a container in which the postman can place the parcel.

The car literally grows into an independent stakeholder—an asset and a companion beyond physical mobility. With all its digital and informative refinements, it offers us comfort, access, security, and experiences—experiences that we normally associate with cinema, a trendy fashion boutique, an art gallery, or a café. Let us think, for example, of the car as a mediator of new products, as a showroom, lounge room, or bedroom. The vehicle becomes a business partner: in the future, we will not only plan, reserve, book, and pay for our experiences in the car but also use the car as a service provider that offers us planning, reservation, booking, and payment services for clothes, products, espresso, and artworks.

Digitalisation, climate change, human innovation and structural challenges also lead to a competitive dispute between OEMs and disruptors. This discourse could evolve even further and occupy much more space in this context if the mobility topics of a city or a region were finally centred in one role, mobility partners aligned according to the required mobility mix of a city, and all relevant tasks bundled into a contextual mobility management system. As things stand today, we can no longer examine the question of sustainable mobility in tiny pieces, only looking at them in isolation from one perspective, but not from the other. Analogous to a corporate strategy, a holistic mobility strategy is required.

In-Car Living: The Car Becoming a Living Space

Digitalisation, climate change, human innovation and structural challenges are forcing the debate about the influence of the car on our living spaces. The car itself becomes a living space that competes increasingly with other living spaces such as the office, home, shops and social venues.

Working with This Book

In the following chapters, we present the concept of the living space of the car from different perspectives, deliberately and in a variety of ways. In parallel, we will raise the question of what role the car will play in the future in our perception and in our actions, becoming either a good or a bad guy in Smart Mobility.

We will go on a journey together. We invite you to consider examples and role model scenarios accompanied by tools, checklists, and project proposals. The depiction of our scenarios takes place partly in fictitious, partly in real, worlds. You will find the roadmap of the book in ◉ Fig. 1.2.

Each chapter concludes with a checklist, giving our readers room for their own thoughts and plans. A bibliography can be found at the end of each chapter. The book also contains a glossary containing essential terms.

Fig. 1.2 Smart Mobility in practice: Multiple dimensions of the automotive habitat

Acknowledgements I would like to thank firstly my reading community that has grown to more than 300.000 in the German speaking and in the international market. It is your drive and mindset that drives us, my interview partners and contributing experts, to serve you with the trends, themes, tools and concepts of smart and inclusive mobility. This book is now continuing the Smart Mobility series that we kicked off several years ago. Use this book as a manual in your daily, strategic, learning, and studying, lives, and your decision taking.

It is your commitment to our efforts that motivate me to research, put my findings and visions into words, illustrate, audiotape and lecture them. Thanks for your support, valuable hints and the trust in my work and vision of inclusive and diverse mobility. The result is a practitioner-focused book that accompanies all those who deal with intelligent mobility, look for smart safety concepts in and around the digitised and independent car, and design the in-car living space as part of their daily business.

Many thanks to Petra Steinmüller from Springer Vieweg and the editorial and production teams for making our work accessible to the world. Many thanks to my team and herein Sarah Tyrer.

Happy reading, DoktorB and team.

References

1. Jakes J (1988) Firebird. A dystopial from 21st century America. German edition. Goldmann Publishing House, Munich
2. Association of the Automotive Industry (2014) Automated driving. https://www.vda.de/de/themen/innovation-und-technik/automatisiertes-fahren/automatisiertes-fahren.html. Accessed 17 Dec 2017

Our Car, the Habitat of Modern Society

2

DoktorB Barbara Flügge and Vinod Jadhav

Abstract

Vehicle, car, transport mean, automobile. Different terms leading us to a human-machine-relationship that has never been stronger than in the present. The present where we reflect upon the relevance of cars for our lives. So, let me ask you how important the car is to you in this moment? We are not here to conduct an isolated study sponsored by the bosses of the automotive industry. It is rather our investigative mindset that holistically connects the car in all its facets with our human desires. Fulfilling our desires seem to be disconnected from the desire to own. The contextual conditions of where we are now, where we work now, and where we learn, is what interests us. We analyse and show how the Persona Setcards™ help you to interact: how the environment, the mechanisms, and the soul of the car and how the car interacts with us, its inhabitants, and visitors. Additionally, the impact on urban and rural areas that move decision-makers and environmentalists.

D. B. Flügge (✉)
digital value creators (DVC), Bottighofen, Switzerland
e-mail: barbara.fluegge@dvcconsult.com

V. Jadhav
SAP SE, Walldorf, Germany
e-mail: v.jadhav@sap.com

B. Flügge (ed.), *Smart Mobility in Practice: Good Car, Bad Car, No Car – Is this the End of Nurturing Our Mobility DNA?*,
https://doi.org/10.1007/978-3-658-43171-6_2

2.1 Knowing Your Position

Barbara Flügge and Vinod Jadhav

On the Importance of Holistic Mobility Examinations

Imagine the efficiency of a mobility management system or a local public transport system that takes care of mobility in cities around the world on a daily basis. What do the timetables look like, or is there on-demand control instead? Which fleet vehicles are due for inspection? What happens if the weather changes and continuous rain is expected, or if the announced road repair is extended by 2 weeks? Will the train crew be on strike in the near future and, if so, will this not only have an effect on increased demand for city buses in terms of transport, but also on the mood of our own colleagues? We can well imagine these and many other questions. But how many of the corresponding answers are shaped by a transport management of the previous century? In reality, certain answers are no longer needed since digitisation takes away some of our analogous considerations and decisions. On the other hand, is there not a need for a new public awareness of innovation to ensure inclusive and interconnected mobility?

The questions that have to be asked are aimed at the importance of the car among the individual population groups. It is about how mobility supply and demand behave in relation to shopping, visits, excursions, concerts, sporting events, health checks and much more! It is no longer sufficient to analyse the number of trips per citizen and the time spent in one's own bus network. Background information such as the above-mentioned mobility behaviour under the influence of one or more events, such as visits and shopping, requires the investigation of the relationship between the use of subway, tram, and car sharing services around the corner and their impact on the observed persona.

The Car in the Context of Urban Mobility

Assuming that intelligent mobility, i.e. smart mobility, is a design element for our living space and does not exist as a necessary, investment-intensive structure, we can turn this line of thought into a comprehensive study. Following natural habitats and ecosystems, every economic and man-made habitat can be mapped as a functioning ecosystem in itself. This includes constant accessibility and mobility for all. An understanding is needed of who holds the reins in modern ecosystems and how. This requires a clear understanding of user groups, providers, decision-makers, and the conditions under which interactions and transactions occur between the parties involved. Questions arise regarding billing and cost-effectiveness and the extent to which mobility services should and can be made available free of charge to subscribers. Consensus is needed to make our living space sustainable. Consensus building is a challenge for many. If the authorities, companies, and interest groups are faced with this task, it is not difficult to develop use cases. These will be covering business modelling, role models and financial feasibility.

In our study, which we have carried out in 20 cities (see Sect. 5.1 [1]), we dived into the significance of the car in urban mobility, applicable rankings, and further city specific mobility projects.

The first study (see ◉ Table 2.1 [1]) deals with the cities of Bangalore, Barcelona, Beijing, Berlin, Guangzhou, Hong Kong, London, Madrid, Melbourne, New York, Seoul, Shanghai, Singapore, Stockholm, Sydney, Taipei, Tokyo, Warsaw, Washington DC, and Vienna. We used features such as fleet size of the public city bus network, length of the road network, traffic density on the road, number of motor vehicles, number of private cars, daily trips by train or daily trips by public bus (◉ Table 2.1).

Cities with a comparatively large geographical area, such as Sydney and Beijing, have a high proportion of private cars. While Sydney—and also Melbourne—are investing in the expansion of the subway network, and their proportion of private cars can be justified by a lack of public transport, the desire for a car in Beijing is still rising steadily despite adequate transport options. This is due to increasing prosperity and its association with being able to drive a car yourself, as the desire to show wealth to the outside world by owning one's own vehicle has increased. Other cities, such as Berlin or Vienna, have a relatively low proportion of private cars. In the case of Vienna, it can be seen that commuters make use of the public mobility services of buses, trains, subways, and trams, which are geographically well coordinated beyond the city limits. Clientele who can afford a parking space as well as a flat in the city centre will still continue to prefer driving their own car, despite the well-developed mobility offering in the city. The car then symbolises a different status and can also be used for representative purposes.

Table 2.1 Criteria for positioning public transport and private cars in 20 cities

Criteria	Attributes				
Share[a] of private cars[b]	0–600	600–1200	1200–1800	1800–2400	2400–6000
	Barcelona Hong Kong Singapore Stockholm Taipei	Berlin Shanghai Vienna Warsaw Washington DC	Bangalore Guanzhou Madrid New York Tokyo	Melbourne Seoul	Beijing Sydney
Number of stations[c]	0–100	100–200	200–300	300–400	400–500
	Stockholm Warsaw Washington DC	Bangalore Barcelona Berlin Guanzhou Hong Kong Singapore Sydney Taipei Vienna	Beijing Madrid Melbourne Tokyo	London Seoul	New York

(continued)

Table 2.1 (continued)

Criteria	Attributes				
Daily trips[d] by train[c]	0–2000	2000–4000	4000–6000	6000–8000	8000–10000
	Barcelona Berlin Madrid Melbourne Stockholm Sydney Taipei Vienna Warsaw Washington DC	London Seoul Singapore	Bangalore Guanzhou Hong Kong New York Shanghai	Beijing	Tokyo
Fleet size public bus	0–5000	5000–10000	10000–15000	15000–20000	20000–25000
	Barcelona Berlin Madrid Melbourne New York Singapore Stockholm Sydney Taipei Tokyo Vienna Warsaw Washington DC	Bangalore Hong Kong London Seoul	Guanzhou	Shanghai	Beijing
Daily trips[d] by public bus	0–3000	3000–6000	6000–9000	9000–12000	12000–15000
	Barcelona Berlin Madrid Melbourne New York Stockholm Sydney Taipei Tokyo Vienna Warsaw Washington DC	Bangalore Hong Kong Seoul Singapore	Guanzhou London Shanghai		Beijing

(continued)

Table 2.1 (continued)

Criteria	Attributes				
Area [km²]	0–4000	4000–8000	8000–12000	12000–16000	16000–20000
	Bangalore Barcelona Berlin Hong Kong London Madrid New York Seoul Singapore Taipei Tokyo Vienna Warsaw Washington DC	Guanzhou Shanghai Stockholm	Melbourne	Beijing Sydney	

[a]Number of vehicles
[b]Private cars in .000
[c]Metro and/or tram
[d]Daily trips in .000

Public metro and rail services such as suburban trains are heavily frequented in major cities and megacities such as New York, Bangalore, Shanghai, Guangzhou, Tokyo, and Beijing. Alternative or supplementary offers by public city buses are then less used in small cities.

In Singapore, private car ownership accounts for around 10% of the population, i.e. around 575,000 vehicles. Compared to other cities, Singapore offers a comprehensive range of mobility services from the public sector. The EZ Link card [3] also contributes to the easy use of buses, metros, and taxis. Nevertheless, and although the administrative and fiscal hurdles [4] are high, targeting middle class vehicle ownership is a declared goal. Regulation by the Land Transport Authority (LTA) in Singapore to limit vehicle registrations for environmental reasons aims to reduce air pollution and traffic congestion. Furthermore, any usable space in Singapore is crucial for further growth—both for residential construction and for the establishment of companies and research facilities. In the meantime, Singapore has frozen the vehicle population and further vehicle registrations have not been permitted since February 2018 [4].

Mobility Analysis I

A demographic representation of the population and income strata is useful to investigate the dependencies between mobility consumption of public transport and ownership of private cars.

User Behaviour in the Mobile Context

With evaluations as mentioned above, we obtained a rough estimate of the mobility offer in cities measured by the number of providers and their use. Roughly because of fleet sizes and usage figures of public bus, metro (metro rail transport, MRT), and tram (light rail transport, LRT) services, as well as the identification of privately available vehicles, in particular passenger cars, figures do not yet provide any information about the usage behaviour in general and the significance of the car in particular. With "usage behaviour" we refer to the decision to accept one or more means of transport as mobility consumers for a certain distance. A further aspect to consider concerns the duration of the distance travelled, and what percentage of such distance is undertaken through the chosen means of transport. Commuter movements differ from visitors, tourists, and business travellers. It is therefore recommended to include commuter movements for consumer goods and goods for daily use.

Mobility Analysis II

The analysis of the travelled distances is an important criterion for the choice or deselection of the car as a means of transport for both passengers and goods.

From our own behaviour we can see that we often choose the car, although this is often when our next stop is in the city or in a certain area in our immediate neighbourhood. Although we aimed to undertake a detailed evaluation of public bus offerings, details were missing, such as the number of bus stops. Identifying the number of public bus stops in an international context, we had to consider semantic distinctions of "bus stop", "bus station", German "Haltestelle" and "Busstation" amongst other terms. Additionally, search results also ended up considering bus central stations. Of course, those are important, too, serving as crucial important intersections and switchboards for commuters and mobility routes.

In some cases, published figures or information on the urban mobility infrastructure are lacking. Nor are there any reliable figures for bus routes and bus stops operated or serviced by private operators across all 20 cities. We found an interactive bus map with all bus stops in the internet portal of the public transport company of Barcelona [5]. In comparison, Berlin has published bus timetables and bus stop overviews sorted by districts and compass directions [6]. An example of a private operator is Flixbus [7] with its services in Germany, Austria, Switzerland, and Italy. Flixbus currently uses public bus stops. It would be conceivable that Flixbus, or similar companies, could set up their own stops at suitable junction traffic points, especially in the regional area. There are also stops for airport shuttles in some of the cities with airport connections. Other examples are new lines and mobile stops, and the so called "mobility bench". These are the result of driving services for senior citizens, such as the free shuttle service in Sydney [8] or the voluntary senior citizens' shuttle in Hohenburg [9]. Other new driving services include shopping buses [8]. Examples of mobility benches and carpool benches can be found in Adelebsen [10] or Lödla [11], Germany. In Lödla, the ride sharing bench was also called an innovative form of hitchhiking!

A further aspect of the importance of the route analysis relates to the 30-min studies of Australian cities. This involves pushing for a maximum time expenditure of 30 min, which we spend on the accessibility of home, place of work and leisure time. Will the so-called 30-min city thus become a design element for modern urban development? Aren't the Kiez idea, as it is lived in cities like Berlin, or the neighbourhood project, such as that of Zurich in Europaallee [12], the same? Maybe we should be considering a 15-min city rather than the 30-min city.

Without digressing too much into urban planning topics, the evaluation of accessibility, affordable mobility and perceived comfort takes place here too, first and foremost independently of the 30-min approach in our minds and habits.

Purpose Driven Decisions

Mobility decisions are earmarked and then self-determined, as long as they fit into our personal optimum of accessibility, affordability and comfort of the mobility offered.

The Ranking Is Not Everything

Some of the survey and Smart Mobility Indices are interlinked with city analyses, such as the Urban Mobility Index 2.0 by Arthur D. Little [13], the Sustainable Cities Mobility Index 2017 by Arcadis [14], and the INRIX Global Traffic Scorecard established in 2017 [15].

Rural areas, on the other hand, are often missed—neglected and forgotten—not only in statistics and studies, but also in wider decision-making. We recommend carrying out studies on rural areas, whilst also keeping the increasing densification and geographical expansion of cities in mind. Rural areas deserve an upgrade and should be taken into account in smart studies! In the course of our research for this book, we have come across individual projects through publications in regional newspapers and through the announcement of citizens' and administrative initiatives.

It remains to be hoped that the efforts contained in Chapter IX, Worth Living Cities, Attractive Regions and Affordable Housing, as contained in the coalition agreement of the Grand Coalition in the Federal Republic of Germany, as of 07 February 2018, will be implemented ([16], p. 109 ff.). Point 7, for example, focuses on passenger transport law, public transport, and mobility in rural areas (ibid., p. 121 ff.).

Making Needs Heard in Rural Areas

Rural areas deserve attention. If traffic collapse, air and noise pollution, and high living costs in urban areas, underpin the conduct and financing of innovative projects in cities, the population and infrastructure requirements in rural regions must not be neglected. It is precisely because of digitisation and virtualisation that conclusive concepts worth developing and living in the countryside can be created and implemented.

The city of Aarhus, with around 262,000 inhabitants [17], ranks itself in the lower segment of medium-sized cities in the above-mentioned study. Compared to Warsaw, with a population of 1.7 million inhabitants [18], and Stockholm, with a population of 900,000 inhabitants and a catchment area of another 1.3 million inhabitants [19], Aarhus is manageable. Nevertheless, it faces similar challenges to other cities when it comes to congestion and excessive infrastructure costs. Aarhus is looking for approaches for infrastructure improvements and a move away from private cars in the city. Existing capacities are to be better utilised and better coordinated with each other; this is in line with Mobility-as-a-Service (MaaS).

Mobility-as-a-Service (MaaS) at a Glance
Mobility-as-a-Service means that the user pays for a portfolio of transport means to access mobility and consume it as a service. She pays per trip or monthly. For example, a user can travel by public passenger bus for an hour, access a taxi to ensure they are on time to their meeting, and get free mileage to use a car rental.

A detailed introduction to MaaS can be found in ([1], pp. 134–141). If existing capacities are better used, this relieves the space situation in the city and leads to less energy demand. In addition, fewer cars mean less air pollution. The following remarks are taken from Susanne Krawack's lecture and presentation at Fluidtime in 2017 in Vienna [20]. Susanne Krawack is the head of the mobility department of the city of Arhus. In addition, as occupancy rates rise, travel by public transport should become less expensive for individual users. The following parameters are in the foreground:

- Introduction of ridesharing, i.e. car pooling
- Establishment of a mobility provider for the operation of ridesharing
- Traffic mix of ridesharing and public transport
- Introduction of an IT system for better information processing, flexible deployment plans for emergency personnel and the introduction of additional services
- Introduction of necessary incentives to encourage citizens to use MaaS

The city administration of Aarhus is consciously looking for a way out through holistic mobility management. If the train services have so far been viewed in isolation by Danish Rail, Ride Share providers and owners of private cars, Aarhus will seek a merger. If Aarhus remains consistent and orchestrates mobility needs across all modes of transport—bus, train, car, bicycle, and mobility routes—Aarhus is well on its way to becoming a model for holistic mobility management! Aarhus launched a piloting of MaaS in 2017. A tender for the implementation of MaaS was due in 2018.

In the cities we examined, there are also examples that on the one hand question the role of the car, but on the other hand at the same time at least partially take into account the role of the car as a MaaS member. In Vienna, the so-called Vienna Technical Concept for 2025 aims at an 80% fulfilment of mobility needs—personal and goods related—via the environmental network [21]. The environmental network includes all means of public transport, as well as travelling on foot or by bicycle. This leaves 20% use for individual transport in 2025. The current level of individual traffic is 28%.

First research results on MaaS can be found in Great Britain [22]. The preview report contains a statement by George Hazel that MaaS is competing with the car. In our opinion, this is true because MaaS optimises the use of existing capacities and reduces the use of private cars—unless they are deliberately used for ridesharing, for example. However, we do not want to share a complete renunciation of the car in the MaaS concept in this way. In our opinion, the transformation of the car away from the established means of transport takes place for a specific purpose. For example, the car will be indispensable as a medium for the intelligent mobility mix in rural areas. Similarly, the desire for ownership will persist, whether for representational purposes, for families, or for citizens who want to afford themselves the flexibility of an alternative medium of mobility.

With the evaluation of the Swiss intermodal Green Class mobility project in Chap. 7, we lead the discourse concerning whether an intermodal offer from a railway operator, a private BMW car owner, and a public-private e-bike network would be sufficient to break with old habits.

Urban projects often only reach the public if cities are aware of the advertising character of their projects and know how to involve the participating organisations accordingly. Another aspect is the assessability and comparability of projects. Being classified and evaluated as a city in global or regional indices helps the projects from two essential points of view: indices and urban comparisons provide (a) argumentation aids for one's own projects and (b) certainty in the assessment of past projects and their effect on one's own position, irrespective of whether the latter aspect has been assessed as positive or inhibitory from the outside.

Nevertheless, municipal decision-makers and project teams like to get lost in rankings and external evaluations. What is more important for the success of the projects are your own assessment of the current situation and the handling of externally acquired knowledge. You will find more about city assessment under chapter Sect. 3.1, driven by the use case of Johannes.

The Human Being Decides

We ourselves trigger and create the transformation options through thought games that open up new ideas for the transformation of the car—moving away from the established means of transport towards new areas of application. Consumer tourism and our quest for omnipresent mobility continue to be at the forefront nevertheless.

2.2 What It Is Worth to Us

Barbara Flügge

Living Spaces
Along with the settlement currents of humanity and the founding of communities, cities, and villages as urban and regional, living spaces are shaped by the coexistence and interdependence of their inhabitants [23]. They are, at the same time, part of economic and political systems. As individuals, we associate the term "living space" with precisely those places that most closely correspond to our personal feelings of security, comfort, growth, and individuality. Examples of this are our living, working and leisure rooms. In doing so, we are interested in the best possible fitting equipment, such as furniture, music, and colour accents. Some worry about an isolated, yet fenced, habitat, others live outside their habitats with the community and invite us to experience and stay together.

In comparison to ecosystems, the quality of a habitat is also oriented "towards the degree of togetherness and reciprocity, access to resources and facilities, and the sensible use and utilisation of the existing habitat" irrespective of size, geography, and social demands ([1], pp. 27–28).

With the invention of the automobile and the progressive liberation from dependencies defined by others, we successively opened up a new living space for ourselves: our car. Our drive for perfection, individuality, and creativity has not stopped at the living space of the car. It has emerged into an addiction. We became dependent in two directions:

1. The development of known and unknown geographies is no longer conceivable without vehicle-based locomotion.
2. Spatial development in cities and regions followed, and still follows, the requirements of vehicle-based locomotion and mobility demand.

Published in the year 2023, Nora s. Stampfl wrote in agora42 about the subject of mobilisation and discusses how men have successively made themselves dependent on "mass motorisation" through technological progress and adaptation [24]. Thus, according to Stampfl, renunciation of the medium of the car would be tantamount to a detoxification cure (ibid).

Let's take a quick look at the expenses we are willing to incur in connection with the car as a means of transport. The Swiss Automobile Club (TCS) calculates an annual expenditure of CHF 10,487 for a sample vehicle with a new price of CHF 35,000, and thus an average cost of CHF 0.70 per km [25]. According to the TCS sample calculation, around 38% of this is attributable to mobile, i.e. mileage-dependent, expenses for fuel, tyres, service, and repairs. This also includes an average impairment of 10%. Fixed costs such as depreciation, interest on capital, traffic tax, liability insurance, partial comprehensive insurance, ancillary expenses, parking costs and vehicle maintenance account for approximately 62%.

Energy costs such as short-term stress and nervousness were not taken into account, nor health effects caused by long journey times, traffic density, danger zones such as construction sites or traffic jams. Nor have we found further monetary expenditures for therapeutic and anti-stress measures. Furthermore, there is no survey of the financial consequences for business and private travellers due to late arrivals, short-term postponements of appointments, missed connections in transit, and continued delays in meeting schedules. Conversely, it should be possible to derive countermeasures from this, such as running education lectures in national and international media, in traffic education with newcomers, by health and health insurance funds, or by means of information in navigation systems and on warning or signal billboards. The 360-degree view sharpens the view:

- Where can remedial action be taken?
- Where does incremental progress take place, or even progressive, radical or disruptive development? [26]

Let's be conscious about positive and negative universal phenomena and the associated cause and effect of decision-making in the innovation process. Are there other industries with analogous effects or dependencies? And what contributions can be made with little effort and resources? These are only two of a number of questions that are applied in the innovation process.

Next Stop: Experience Spaces

In this volume, we invite you into the discourse on the significance of the car in urban and rural mobility. Think about the discontinuation of bus stops for public bus lines, which are dismantled by the use of autonomous vehicles. The latter offer a flexible route design in the future and promise a drastic reduction in waiting times. Through intelligent on-demand control of the desired route, being determined from our location, our personal experience factor and the destination, we will be served ad hoc and without pre-booking vehicles or studying bus schedules. That's the vision.

It remains to be seen to what extent the use of autonomous vehicles will bring public bus transport to a standstill. Measured by the usage figures in ◉ Table 2.1 with several hundred thousand trips per day in medium-sized cities and up to 10 million and more trips in cities such as Beijing, a replacement of the city bus is unrealistic.

The extent to which bus routes can be replaced by autonomous vehicles in the future, for example, requires a more detailed reappraisal of the role of city buses than is currently the case. In addition to the number of buses, timetables, and lines, this includes the number of stops and the orientation of intermodal nodes. The latter provide connections for intercity buses, underground trains, and trams. It is necessary to study the intermodal mix with the addition of bicycle routes and bicycle stations to change from bus to, or docking to, e-bike routes.

The degree of utilisation over different time spans, such as time, day, month, and season, are also included. In addition, we evaluated the smooth transition from one medium to the next according to our own security, but also the security of our assets such as bicycles, accessories, and purchases.

The autonomous vehicle not only picks us up and takes us to our destination. It lets us consume experiences while driving. What if the digital assistant projects our wishes and hotspots, translates them into operations, and serves our needs. In this case, we would already have the experience at hand in person-controlled vehicles. We would take the advantage from intermodal mobility through carsharing and ridesharing providers. Who says it needs autonomous vehicles?

What have we learnt so far? It is we who are shaping the world of experience for ourselves and for our customers. We acknowledge the impact because we are guiding, not only being guided, in an experiential journey of mobility. We acknowledge the experience and its after-the-fact intensity even when the mobility moment is gone. See Chap. 4.

Does the Mobility Provider Strive for Gifting Us a Guided Experience or the Past Perceived Experience?

In the future, experiential mobility will determine which means of transport and which degree of automation we prefer for certain routes. Mobility that feels like an experience will determine whether we actually use that means of transport again.

The Omnipresent Mobility

Today we expect an omnipresent mobility. We expect to be anywhere in the world at any time. Information technology and the art of engineering have made a corresponding contribution to this in recent decades. Being omnipresent also means being able to move and decide freely.

If we are free in our decisions—with regards to aspects such as affordability and comfort—and if the expected flexibility is met, it will be easier for us to switch to alternative mobility offers. Easier because we receive an adequate replacement service. Otherwise, we wouldn't be able to do without the car without something in return, although paradoxically—and we know this from ourselves—we can foresee that we would be stuck in traffic, arriving late and annoyed. The car has increasingly developed into a habitat for us.

> The former Secretary of Transportation of the United States of America, Secretary Anthony Foxx, put it this way: "Transportation is not just about concrete and steel. It's about how people want to live." ([27], p. 2)

In other words, transport is less about the actual mobility structure and more about how people experience mobility and whether they believe mobility modes can meet their expectations. Additionally, we are talking about habitats that we have carefully conquered, built, and expanded in recent years. And we don't like to give that up. Our bond(s) as car owners to the vehicle are much too close, as shown in ◉ Fig. 2.1.

Even passengers can be motivated by participating in the car system. There are one or more possible reasons for this:

- Social commitment: The car gives us access to communities and the preservation of a community.
- Value-driven binding: The car is responsible for preserving and communicating cultural values. It strengthens our value system and world views concerning wealth, privilege and status, and conveys these values not only in its use but simply in the statement of ownership.
- Commercial bonding: The car guarantees our income. We need it when conducting an economic activity in the service field or the exercise of services. It transports, provides, stores, and supplies the economic cycle with its own and borrowed services.
- Purpose driven commitment: Without a car, earmarked collaborations or temporary alliances, for example for attending an event, a trade fair, or a conference, are often inconceivable. This includes a city visit with family, a class reunion or project coordination with colleagues on the way to the customer's appointment.

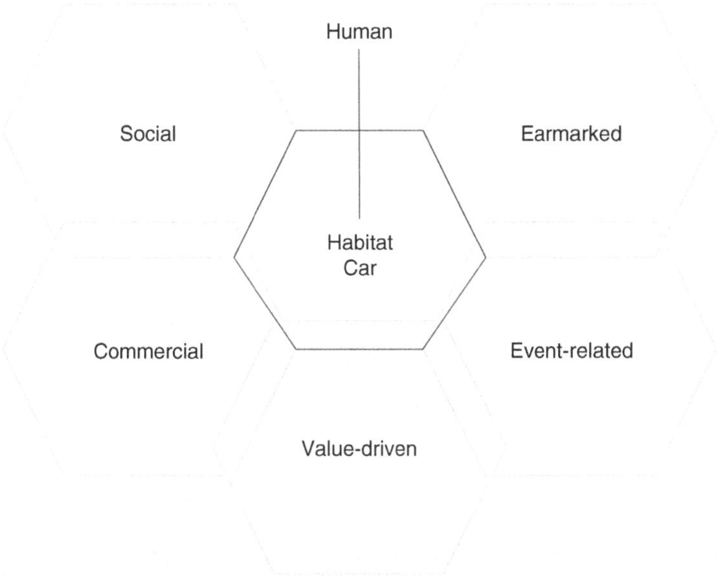

Fig. 2.1 Weaving the bonds between humans and cars

- Incident enhanced commitment: Ensuring that we can or must have access to a vehicle in the event of an emergency or other occurrence requiring immediate action gives us a sense of wellbeing; be it to quickly fetch the forgotten gift for grandma again, or the short-term emergency visit to the doctor, being provided with the mobile care service, or even the convenient purchase of food via so-called potato taxis.

But how do cities deal with the car as the living space we have fought so hard for and the persistence of their travelling population? Particularly in US cities and long journeys through large catchment areas, it is worth looking at the results of the Smart City Challenge, which was launched as early as 2015 by the United States Department of Transportation (USDOT) as part of the Beyond Traffic 2045 master plan [27]. In the competition, USDOT assessed how cities plan to ensure the sustainable mobility of people and goods by means of infrastructure, innovation, and digitisation measures. The call was answered by 78 cities. The seven finalists included the cities of Austin, Columbus, Denver, Kansas City, Pittsburgh, Portland, and San Francisco. The finalists received a budget of USD 100,000 for a 3-month project phase, marketing support and the support of experts from both public and private institutions. The finalists' plans contain valuable information on how to deal with the car as a habitat.

The main influencing factors are presented in simplified form in ◉ Fig. 2.2. The illustration is easily usable for, or referable in, our own projects. Contact us at https://www.barbara-fluegge.com/en-contact to get the diagram and evaluate the influencing factors for your project.

The proposed means of transport may be adapted according to their relevance to the purpose of life or transport. For the evaluation, which location you consider plays an important role.

It is no longer just a matter of a location such as a city or a place of work. Through geographical expansion, the range of services expands, too. Further rooms such as event rooms and mobile or capsule accommodations will be added. Depending on our preference, some influencing factors are less or more important for the consumer (Sect. 3.3), such as the commuter who opts for the e-mobility scooter (◉ Fig. 2.3).

The discourse about mobility providers and users shows the following two poles of interest:

1. The local council defines the mobility mix on the basis of its infrastructural and financial possibilities, political influence and the expectations of the citizens.
2. On the other hand, there is the expectation of a business traveller who passes the district on his way by car.

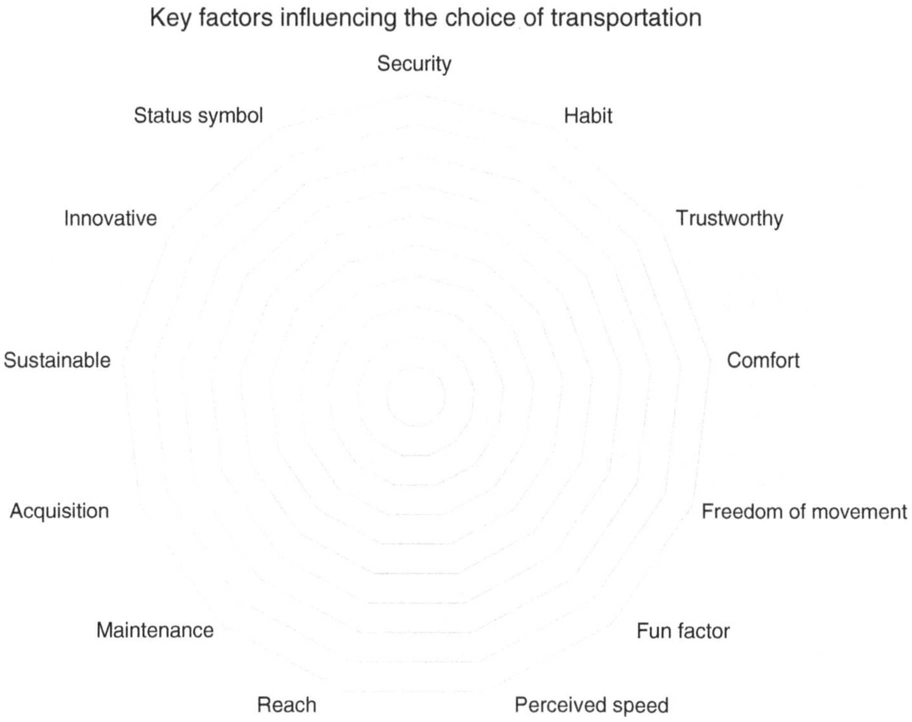

Fig. 2.2 Factors influencing our trust in a particular mode of transport

The Claim Attitude in the Sign of Digitisation

Our demand for omnipresent mobility has grown with increasing technological progress and globalisation. We expect to be mobile at all times: a departure is impossible. More than ever, network-based innovation management and innovators are needed to help us bring demand-driven offerings to market.

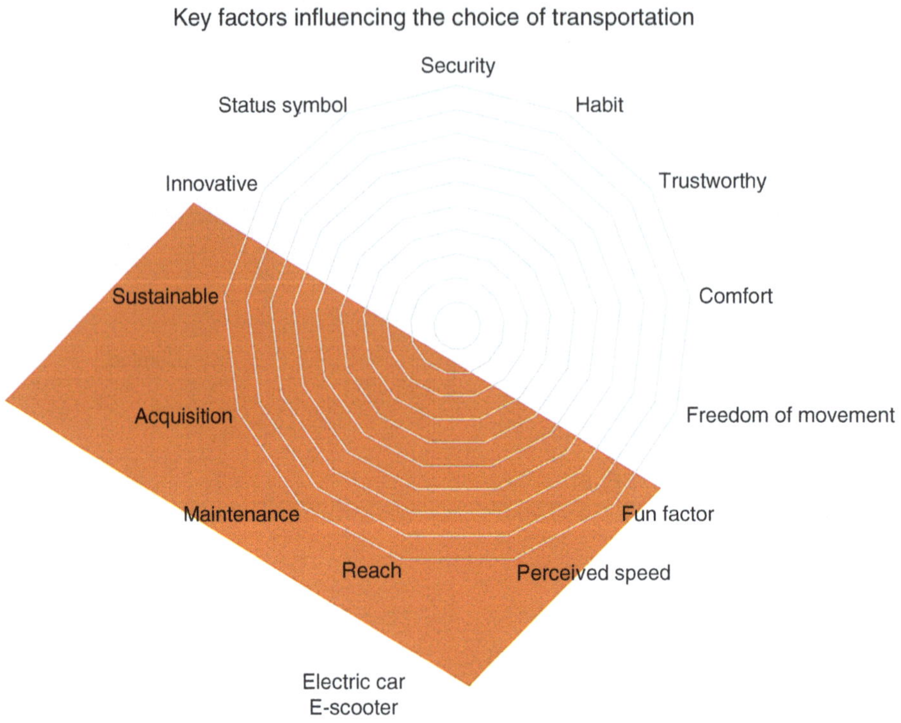

Fig. 2.3 Person-related influencing factors that lead to a changeover to electric cars or electric scooters

2.3 Connecting Everyone

Barbara Flügge

The Forgotten Users

With increased demands for driving bans on diesel vehicles and increased visibility in rankings of congestion and air and noise pollution, in order to tackle the issue all political decision-makers would need to strongly consider the current situation, otherwise it could become too late. In 2017, for example, car drivers spent 51 h in traffic jams in a city like Munich [28]. The increasingly public discussion also shows, however, that public transport operators are called upon to analyse and orient the use of public transport in a more targeted and user-oriented way. It is about the needs of employers and employees, families and leisure enthusiasts, visitors and products that find their buyers and suppliers in the city. This is the only way to successfully find suitable offers and personalised services in order to break down behavioural patterns and inspire people to use alternative offers.

Smart new mobility will be based on the interplay of known needs and changes in needs. The latter results from external factors: increased cost of living, location-dependent components such as digital training, further education and job offers, and user-friendly offers for families, single parents, children, and senior citizens. In the job market, more flexible working time models in particular are leading to an expansion of the geographical area of deployment and the associated change in the mobilisation of employees. Among other things, it addresses sales representatives, freelancers, consultants, software developers and occupational groups with a focus on technical, craft, and creative skills. In the future, ideas will be conveyed to travellers on the move. The variety of topics ranges from news, knowledge transfer, fashion articles, updates in trade magazines, music, videos, shopping tips, training, the mediation of new contacts, and the maintenance of relevant networks such as LinkedIn and WeChat to editorial articles. The latter are prepared and offered for Medium, Twitter, your own blog and WhatsApp according to the duration of your trip and stay. However, an offer of this kind does not require an autonomous vehicle. A needs-based treatment can be carried out in public transport as well as in shared private cars or taxis and company shuttles.

If we ask ourselves about the significance of the car in our everyday lives, we become aware of our own demands and behaviour—our habits. We imagine expected limitations if we were tomorrow without a car. By turning away from cars, however, we benefit from financial savings such as the elimination of maintenance and care, the cancellation of car insurance, the release from the hectic search for a parking space and a new sense of calm for the morning start to everyday office life. At best, it will be cheaper and more relaxed, without question.

Now imagine the construction worker or the postman in the country. How would your colleagues feel without a car? If we had our families in mind, we would include the needs of all, primarily financial and qualitative criteria, in the list of advantages and disadvantages. When analysing user needs, do we therefore prefer city dwellers to the percentage distribution of urban population to rural population?

How do automobile companies do this—do they run user-centric and persona-relevant studies? And how do public mobility providers proceed? Or is it a question of willpower and way-power to take disruptive action and to question previous processes? Start-ups in the mobility segment can be expected to adopt a resilience-enhanced design thinking approach: thus, resiliency becomes part of the methods and offerings to support associations, interest groups, and universities and municipal institutions. Bootcamps and start-up conferences would be inconceivable without workshops that are influenced by design thinking. With few on-board resources, and even less personnel, much can be achieved with little effort. Interestingly, newcomers use procedures that have long been used in consumer goods marketing; see Leitherer [29] and Schweiger [30].

We recommend conducting a requirement analysis driven by the user groups in your Smart Mobility Ecosystem:

- Creating a big picture of the smart mobility ecosystem by evaluating the tasks of all user groups involved (see [1], p. 82)
- Assessing the ties between the user groups and herein distilling what requirements are needed and how often ([1], pp. 80–81).
- Gaining clarity about the core requirements, so that you keep focus on the top priorities in your ecosystem

Persona Relevant and Purpose Driven Requirements Analysis Including Interconnectedness
The purpose driven analysis in the context of mobility leads to a more precise picture as to why someone should actually be provided with which services by which providers and mobility means. This creates networking opportunities between user groups through short relations mapping. Both actions are important because individual job and task profiles are often similar in their needs, even if they are named or described differently and have been moved by completely different means of transport in the past. In some publications there was little detail about personas. Other studies are lacking in their diversity of addressees and simply focus on primary mobility consumers, namely commuters and business travellers and those that use publicly served trains and buses. What we have added are those who travel to the place of work, for example to serve the train, to work as a conductor, or to repair or clean the train.

Let's now take a look at how we can carry out the assessment of mobility and service needs on a continuous basis, applying the concept of the persona. A persona "describes [...] an archetype that is composed of similar or concurring consumer behavior" ([1], p. 45). This includes the commuter, a business traveller, the family doctor, or the taxi driver, for example.

Service providers use a persona to describe exactly who they want to offer which service to. Building on this, further business modelling processes such as service design, pricing, reach analysis, marketing conception and acceptance modelling take place. Personae interact and discuss, they exchange and their paths cross not only in the sense of mobility, but for their short-, medium-, and long-term career and leisure goals. Through their consumption and service behaviour, they belong "to the actual designers of an ecosystem" ([1], p. 46). The more concrete its description is and the clearer its role in and contribution to the ecosystem mobility is, the persona becomes a central building block of a successful market launch.

The personae who are active in the Smart Mobility Ecosystem were introduced comprehensively under ([1], pp. 68–72), in which they were brought together for the first time on the subject of mobility ([1], p. 72). One facet of the persona is the stakeholder. Stakeholders are those who support a project, are doomed to failure, or can exert a decisive influence. They do not always appear in project plans and remain hidden in the background. Project teams are often content with the well-known stakeholders and ignore those who could assist a project and make it known to a wider audience; they might underestimate those who can risk a project. In the case of the latter, critical voices and negative attitudes are not accounted to the project itself, but claimed to be a result of circumstance or an event of the past.

Each person a represents a dedicated target group whose needs can be specified, extrapolated and predicted today—through the help of digital tools and means. Insights into traffic-time-correlations, in- and outbound traffic of service staff, movement of products from Amazon's and third party warehouses, alongside packaging material, as well as insights into the ratio of digital and onsite purchases, are helpful measures to determine the service and consumer behaviour in your city. A further two measures that we would like to point out are the duration of stay and the degree of settlement.

An example: The personas that shape a city's infrastructure, for example, can be determined in conjunction with planned residential homes, the expansion of existing land areas, and the adaptation of the land-use plan. A further indication is the conspicuous increase in demand for building materials, particularly in one target area, a high frequency of do-it-yourself markets, and an increased demand for rental vehicles in the upper load segment.

However, digital insights into how individual personae, for example a young family with its first child, behave is largely manual analytical work. It requires more steps and time, such as intelligent apps and online searches, in conjunction with social media, interviews and observations, to depict user-focused or behavioural-derived notable details. We observe, for example, Twitter behaviour on the subject of home improvement, cries for help such as "How do I do this?" and "Which craftsmen are the best in my area?", as well as increased views of interior design and do-it-yourself videos on YouTube.

Furthermore, the interconnectedness between one's own behaviour and that of third parties is more manually than digitally advanced. The relevance of a persona analysis has been concentrated to marketing and focuses on a few characteristics such as age group, gender, income class, plus rural or urban mobility. Our digital life in social networks and our input concerning ratings and recommendations in searching for and making bookings has helped refine the offerings from product and service providers. In return, our expectations have grown and we have been able to determine the following trend results: (1) there is a growing and progressive individualisation of the consumer offer and (2) a progressive demand for personal address and fulfilment of our needs. Paradoxically, we have become so accustomed to it that even with all anonymity on the internet, we want to be served individually and entirely from start-to-finish according to our wishes.

Knowing Your Service Consumer

We record each persona through the Persona Setcard™. Persona Setcards™ are all-rounders that facilitate person-oriented research, which is indispensable in the context of digitisation projects. We have compiled the most important criteria for integrated persona profiling and put all of our findings in a box of 44 Persona Setcards™. Each Persona Setcard™ provides orientation for the scope of the project and provides information on essential functions and processes. It serves along the scrum-phases as a test basis for development teams. Especially with distributed teams, coordination of people and tasks based on the Persona Setcards™ is essential for team success. The Persona Setcards™ become the common thread.

In the test phase, you set-up user interviews based on the Persona Setcards™ and carry them out together. Interview candidates respond to your scoping ideas and exchange with you on pricing in a relaxed, practical manner.

In the exploratory phase of your project with the Persona Setcards™, project teams collect ideas and obtain internal and external inspiration. Together with the preference analysis, you save time and gain concrete insights (see Sect. 12.3).

You're going to analyse at least ten people, therefore filling in ten Persona Setcards™. Check the extent to which these actually fit in with your project in terms of location—city or country. Do the persons find themselves in your mobility goals? Are there any examples of personae in your area? Or are you already thinking about the next stage of your mobility project? If you want to count a new persona from the entered personae among your customers in the future, make a note of this and mark the set cards accordingly. If known persons are found among the set cards, take the trouble, and evaluate them on the basis of your previous offers and target group assessments. Each Persona Setcard™ offers enough space for adjustments and comments!

Have you thought of a Persona Setcard™ for the means of transport? Use the Persona Setcards™ for the city bus and the ride sharing bench [10]. Means of transport are service consumers, the same as human consumers. Any item, thing, and means is subject to service engineering. Consider the car in the city as obsolete, think of car rental companies, car wash operators or taxi drivers, spare parts suppliers, and use the time to create a Persona Setcard for the car as part of Mobility as a Service. Contact us to receive your package of Persona Setcards™ and the online training course.

The Persona Setcards™: Persona-Driven Value Analysis

Use the Persona Setcards™ and create at least ten Persona Setcards™ for your project. Be accurate and check with the people in your area, your project group, or the competition in the neighbouring town, that you have captured all the details well. If necessary, add additions to the setcard.

Exponential Digital Diversity

The accuracy and precision of demand assessments and their description grows with digitalisation. A coherent recording is needed in order to accurately describe mobility offers in a dedicated manner and to digitally process such offers accordingly. In the servitisation segment, as in the consumer goods segment, personalised offerings will become even more popular than we have been able to imagine so far (cf. course segment week 3, unit 2 Responsible Production & Consumption in [31]).

By demanding even more individuality, mobility, and service providers need to outperform and co-create faster. We are also strengthening the digitisation mandate to process adjacent mobility activities, which include design, production, and assembly, as well as mobility services. The W.I.R.E. Think Tanks for Economics, Science and Society in Zurich calculated in 2010 that there are potentially 10^{31} car configuration variants, with a study that investigated the diversity and sustainability of such elements. Studies were carried out on the diversity of variations [32]. Should we, in view of this 10^{31} figure, consider the different equipment variants for a vehicle, as examined by W.I.R.E., or instead focus more on one variant and focus on the resolution of mobility challenges (see [32], p. 137)?

Once the course has been set to digitise, tens of thousands of variants in the car and outside are possible. Driven by desire and demand, services can be processed in fraction of a second. They can be devised depending on the time, weather, mood of the traveller, or their interests, and published within seconds at the preferred digital and/or physical location and event. In previous years, the term "placing in the network" was coined and used. This term has long since become obsolete. It takes little imagination today to visualise how we will be able to access information and offers in the future. Intelligent lightweight processing media such as Smart Watches, intelligent textiles and bracelets have long since found their way into our everyday planning routine. In the future, we will book using voice, eye contact or simply walking and passing by an identification medium. Think about a biometric identification; think about a featured identifier at the kiosk system at train stations, in the city, on the street, at the shopping cart, or on the intelligent outer skin of a car and building.

Let's not forget that the omnipresence of accessibility challenges us to cope with the booming and blooming individualism.

It remains to be seen whether Alexa [33] will cause a sensory overload at home or on the road. In the case of digitisation enthusiasts like this, they need common sense and appropriate control mechanisms to save us from total digital confusion.

Knowing Needs as a Basic Prerequisite for Successful Networking

Preciseness and on time delivery are two key factors in physical mobility (logistics). In a hyper-connected world, digital and virtual mobility are equally important. This is facilitated through the triangulation of the purpose assessment, the Persona Setcards™ based user analysis, as well as relations mapping.

2.4 DoktorB's Checklist

Important Basic Considerations

- How do you assess the mobility situation in your city?
- Which projects do you notice at your location?
- How often do you change your means of transport in a typical work/leisure/event day?
- Why are you using the car?
- What advantages does the car offer you/your team/your colleagues?
- Run a Persona Setcard™ analysis: What are the top 10 personae in your company?
- Which tasks do you do today in the car, which do you want to do tomorrow on the road?
- How do you calculate the significance of the car from the company's point of view?
- How high are your mobility expenses in correlation to the sales expenses in the company?
- What is the mobility mix for the user groups you favour? Compare the combinations of 1-out-of-4 and 1-out-of-3 as described in Chap. 3.

References

1. Flügge B (ed) (2017) Smart mobility – trends, concepts, best practices for intelligent mobility. Springer Vieweg, Wiesbaden
2. Hyve (2017) Telekom ParcelButler. https://www.hyve.net/de/work/references/telekom-paketbutler/. Accessed 17 Dec 2017
3. Wikipedia (2017) EZ-link map. https://de.wikipedia.org/wiki/EZ-link-Karte. Accessed 17 Dec 2017
4. Website Bloomberg (2017) Singapore wants to stop increasing car numbers from February 2018. https://www.bloomberg.com/news/articles/2017-10-23/singapore-to-stop-adding-cars-on-its-roads-from-february-2018. Accessed 17 Dec 2017
5. Transports Metropolitans Barcelona (2017) Barcelona bus map. https://www.tmb.cat/en/barcelona-transport/map/bus. Accessed 17 Dec 2017
6. Berliner Verkehrsgesellschaft BVG (2017) Bus network Marzahn-Hellersdorf from Berlin. http://www.berlinfahrplan.de/pdf/Busnetz_Marzhell_2017-Dezember.pdf. Accessed 17 Dec 2017
7. Website Flixbus (2018) Company website. https://www.flixbus.de/?wt_eid=2151897122300261927&wt_t=1518971227980&_ga=2.218568476.44321116.1518971225-609578413.1518971225&_gac=1.250937650.1518971225.EAIaIQobChMIhpn66PCv2QIVRyjTCh0ytwtPEAAYASAAEgK_TPD_BwE. Accessed 17 Dec 2017
8. Website City of Sydney (2017) Community transport. http://www.cityofsydney.nsw.gov.au/explore/getting-around/community-transport. Accessed 17 Dec 2017

9. Mittelbayerische Zeitung (2017) Handover: Hohenburger have a new senior shuttle. http://www.mittelbayerische.de/region/amberg/hohenburger-haben-neues-senioren-shuttle-20846-art1572312.html. Accessed 17 Dec 2017
10. NDR.DE (2017) The Carpool Bank should contribute to mobility. https://www.ndr.de/nachrichten/niedersachsen/braunschweig_harz_goettingen/Mitfahrerbank-soll-zur-Mobilitaet-beitragen,mitfahrerbank120.html. Accessed 17 Dec 2017
11. Leipziger Volkszeitung (2017) Lödla revolutionises hitchhiking in the countryside. http://www.lvz.de/Region/Altenburg/Loedla-revolutioniert-das-Trampen-auf-dem-Lande. Accessed 17 Dec 2017
12. SBB (2016) Europaallee – building site H. Guidelines for retail space applications. http://marktlagerstrasse.ch/sites/default/files/plot_files/Leitfaden%20Retail%20Baufeld%20H_160906.pdf. Accessed 17 Dec 2017
13. UITP and Arthur D. Little (2014) The future of urban mobility 2.0. http://www.uitp.org/sites/default/files/members/140124%20Arthur%20D.%20Little%20%26%20UITP_Future%20of%20Urban%20Mobility%202%200_Full%20study.pdf. Accessed 17 Dec 2017
14. Arcadis (2017) Sustainable cities mobility index 2017 – bold moves. https://www.arcadis.com/en/global/our-perspectives/sustainable-cities-mobility-index-2017/. Accessed 17 Dec 2017
15. Website INRIX (2017) INRIX global traffic scorecard. http://inrix.com/scorecard/. Accessed 17 Dec 2017
16. Faz.de (2018) Draft KoaV. A new dawn for Europe – a new dynamic for Germany – a new cohesion for our country. Coalition agreement between CDU, CSU and SPD. http://dynamic.faz.net/download/2018/koalitionsvertrag.pdf. Accessed 17 Dec 2017
17. Population.City (2017) Arhus. . http://population.city/denmark/arhus/. Accessed 17 Feb 2018
18. Population.City (2017) Warsaw. http://population.city/poland/warsaw/. Accessed 17 Feb 2018
19. Population.City (2017) Stockholm. http://worldpopulationreview.com/world-cities/stockholm-population/. Accessed 17 Feb 2018
20. Fluidtime (2017) Fluidtime symposium 2017. https://www.fluidtime.com/symposium/symposium-2017/
21. City of Vienna (2017) The concept of mobility – an overview. https://www.wien.gv.at/stadtentwicklung/strategien/step/step2025/fachkonzepte/mobilitaet/ueberblick.html. Accessed 17 Feb 2018
22. landor Links (2017) Mobility as a service. First annual survey of MaaS in the UK – July 2017. http://www.landor.co.uk/downloads/MaaS_Preview.pdf. Accessed 17 Feb 2018
23. WBGU – German Advisory Council on Global Change (2016) Moving mankind: the transformative power of cities. WBGU, Berlin
24. Stampfl NS (2013) Rolling contradiction – the automobile society on the way to the furious standstill in agora42. https://www.f-21.de/downloads/f21_agora42_Rollender%20Widerspruch_Stampfl.pdf. Accessed 17 Feb 2018
25. TCS (2018) Mileage costs – what does my car cost? https://www.tcs.ch/de/testberichte-ratgeber/ratgeber/kontrollen-unterhalt/kilometerkosten.php. Accessed 17 Feb 2018
26. Lercher H (2017) Big picture the Graz innovation model (Big picture the innovation model). Kleine Zeitung GmbH & Co KG Published by Graz, ISBN 978-3-902819-77-2. https://ssrn.com/abstract=2929258. Accessed 17 Feb 2018
27. U.S. Department of Transportation. Smart city challenge lessons learned. https://www.transportation.gov/sites/dot.gov/files/docs/Smart%20City%20Challenge%20Lessons%20Learned.pdf. Accessed 17 Feb 2018
28. INRIX (2017) Global traffic scorecard. http://inrix.com/scorecard/. Accessed 17 Feb 2018

29. Leitherer E (1971) Demand as a central yield factor. In: Riebel P (ed) Contributions to business economics. VS Publishing House for Social Sciences, Wiesbaden
30. Schweiger G (1975) Determination and weighting of target persons. In: Media selection – data and models. Operation and market (study series). Gabler Publishing House, Wiesbaden
31. Schmid D, Ritzrau W, Flügge B (2017) Sustainability through digital transformation in openSAP. https://open.sap.com/course/sbi2. Accessed 17 Feb 2018
32. W.I.R.E Web for Interdisciplinary Research & Expertise (2010) DOMINO. Handbook for a sustainable world. https://www.thewire.ch/de/publications/domino. Accessed 17 Feb 2018
33. Echo-Tests.de (2018) Alexa – who or what is Alexa and what can she do? https://www.echo-tests.de/alexa-wer-oder-was-ist-alexa-und-was-kann-sie/. Accessed 17 Feb 2018

Growth Areas

3

DoktorB Barbara Flügge

Abstract

Growth areas are geographical (physical) and virtual locations and spaces aiming for growth through existing structural features and a clearly defined execution strategy. Using the example of a medium-sized city in this chapter, we will demonstrate how Mayor Johannes and his project team deal with the growth area in their city. Applying their structured but creative approach, readers will learn about the role of citizen participation and how the car finds its place.

3.1 Growth Creates Opportunities

Growth areas are geographical (physical) and virtual locations and spaces aiming for growth by utilising existing structural features and a clearly defined execution strategy. Examples of growth areas include cities, metropolitan regions, production sites and business networks. We refer to measures of growth in the present chapter as follows: Growth is measured by the successful digital and physical settlement of companies, knowledge and competencies. In growth areas, the safeguarding of their structural and cultural heritage is important, as is positioning themselves attractively for the migration of population groups and investors.

Digitally networked and linked by means of a common strategy, we hereby introduce the so-called Gemeinsam in die Zukunft!, (Into the future together). Joined by this initiative,

D. B. Flügge (✉)
digital value creators (DVC), Bottighofen, Switzerland
e-mail: barbara.fluegge@dvcconsult.com

B. Flügge (ed.), *Smart Mobility in Practice: Good Car, Bad Car, No Car – Is this the End of Nurturing Our Mobility DNA?*,
https://doi.org/10.1007/978-3-658-43171-6_3

the participating German and Polish cities aim to strengthen their socio-economic, communicative and administrative ties [1, 2]. Combined in this way, an area of 160,000 km^2 is considered, with an overall population of 21 million. The concept is based on five key streams:

1. Structural distribution: the use of the advantages of the polycentric structure of small, medium-sized, and large cities, in addition to Berlin and Warsaw
2. Holistic mobility management: improving transport connections
3. Promotion of knowledge building and competence development: Investment programmes in the field of education, training, and research
4. Sustainability: Promotion of sustainable energies taking into account a variety of existing and planned energy producers such as open pit coal mining, offshore wind farms and LNG terminals.
5. Ensuring a high quality of life: the use of scenic and cultural resources, amongst other things, to minimise the risk of natural disasters such as flooding, to increase gross domestic product (GDP) and attractiveness for tourists and locals and, last but not least, to prevent migration and emigration.

Cities and their neighbourhoods aim to sharpen their focus on growth areas by following the five streams and the overall initiative plan behind it. Take a look at your local community together with your partners, i.e. companies, schools, shops, medical practices and the local tourism association, volunteers and voluntary organisations. The cities of Berlin, Breslau, Dresden, Gruenberg, Landsberg an der Warthe, Leipzig, Posen, Potsdam, Schwerin and Stettin are facing up to the challenge in the above-mentioned project. The claim of the initiative thus goes far beyond that of an economic area. It targets the identification of administrative, economic and structural-functional commonalities.

Run the assessment following the outline in ◉ Table 3.1. As soon as the first assessment is available, we will identify what activities are essential for you.

Table 3.1 Exploring the growth space of a city 31[a]

5-point plan	Plan objectives
Structural distribution	Development of a polycentric diagram
Holistic mobility management	Development of a transport connection network following the "public—private—shared" approach
Promotion of knowledge building and competence development	Investment programmes in the field of education, training, and research
Sustainability	Elaboration of the energy demand under consideration of local and distributed energy producers
Securing a high quality of life	Use scenic and cultural resources, minimise the risk of natural disasters, increase GDP and attractiveness for tourists and locals and, last but not least, prevent emigration and migration

[a]The best way to assess the relevance for your city is to use the attached checklist at the end of this chapter

The ◉ Fig. 3.1 also shows the interconnectedness of each of the objectives and that each decision for one or more points from the 5-point plan does not run in isolation, but also provides you with a basis for further activities. The numbering used under ◉ Fig. 3.1 refers to the following activities:

1. Sustainable space and habitat planning
2. Optimised infrastructure and revitalisation of assets
3. Development of a mobility profit and loss accounting and optimised routing
4. Development of an investment programme to take the location into account in tenders, including mobility projects
5. Traffic safety and space safety using modern technologies
6. Secure your location with attractive offers
7. Use of analysis and comparison methods
8. Improvement of individual and freight traffic operations
9. Project planning and capacity build for particularly critical or risky buildings, industries and neighbourhoods.
10. Development of a service offering that is oriented to the needs of the population and target persons.
11. Development and expansion of skills and competences in the field of innovation and digitisation
12. Image and innovation strategies for the municipality

The listed activities are the result of a market analysis with a focus on infrastructure and mobility planning. This shows the character of the networked approach. You don't want to stop at a project, but you want to consistently carry out your projects and master the dependencies as well as possible!

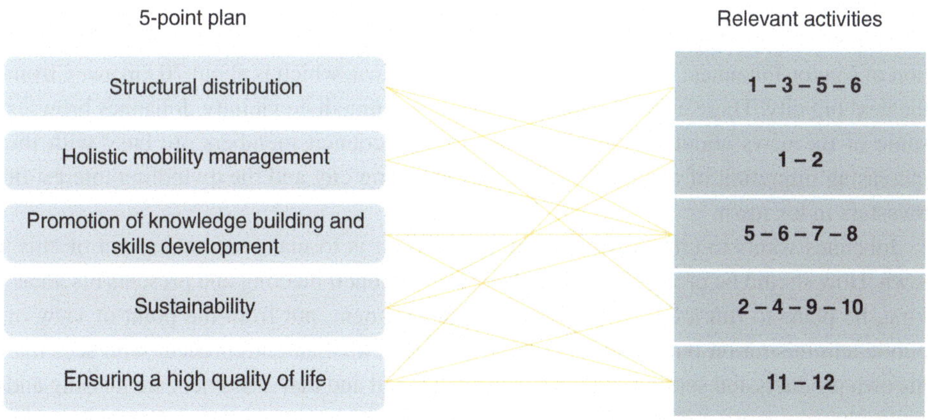

Fig. 3.1 On the implications of strategic decisions on the location of public administration and industry

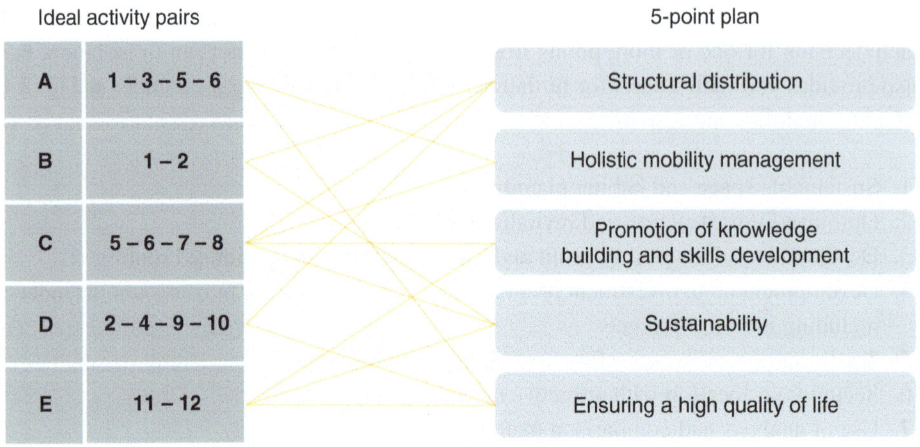

Fig. 3.2 Project planning for Smart Mobility and for Smart City relevant projects

Who takes over which roles? The complexity of mobility management addresses target groups such as product and service providers as well as mobility users and consumers. This sounds obvious. But what the implementation looks like and where to start is a challenge of its own.

In order to provide a viable approach especially for smaller locations, cities, and municipalities or event planners, such as the domestic multi-purpose hall, we have formed so-called activity pairs ◉ Fig. 3.2. Depending on capacity and available funds, you can start with A or B, C, D or E!

3.2 Start Here

Many growth opportunities never make it to the surface. Let us put ourselves in the position of Mayor Johannes. He is the mayor of a small town, which is about 70 km away from the next big city. There are other small towns in the immediate vicinity. Johannes browses some of the news about mobility. He and the city council members are busy with the increasing migration of companies to the neighbouring city and the dwindling interest of investors in his town.

Johannes wants to change something. His interest is to make something out of "his" town. How should he proceed? He schedules a city council meeting and presents his ideas. First, he plans to run a city's location profile assessment, not from the point of view of public administration but rather from the point of view of an entrepreneur who sees that his own products and services are no longer in demand and that customers are missing and prefer the competition in the neighbourhood. Johannes is aware of the challenge faced by his city that originates in a lack of budget to foster growth within the city limits—this growth cannot happen without the urgently needed revenue streams and additional constituent services. Densification in the city centre and the lack of modernisation are two other concerns.

Once the decision has evolved, the city council makes some efforts to advertise a new project. The project's objective is to work on a better image for the city and a better range of services for local residents and employees. In the project, Johannes is joined by two representatives of a local community initiative: the local bakery and the owner of the local physiopractice who resides in the retirement home, as well as two students who moved to the city because of the low rent.

First, the team runs a city profile assessment. Too much has been discussed about what has been neglected in the past. Too often unsuccessful initiatives have been blamed on one of the other local politicians. Johannes and his colleagues keep focus on their questions and sort out the essential elements of their mobility strategy. How they start is depicted in ◉ Fig. 3.3.

The site analysis chart is now the starting point. The team understands that it takes an innovative approach. The team has not yet identified the cause and effect of the underlying challenges of the city and its current status quo. There is speculation about the attractive job offers in other locations and the proximity of other cities. At least from today's perspective, both factors cannot be influenced or changed. Therefore, the team applies the five steps of the Explorative Phase; see ◉ Fig. 3.4 of our Smart Mobility procedure model.

Johannes and his colleagues are putting the situation in the city into concrete terms. For this purpose, each team member describes his or her view of what is happening in the city.

Fig. 3.3 Site analysis for inclusive Smart Mobility projects

		Completion degree evaluated by the project team by Johannes	Completion degree evaluated by the involved citizens	Completion degree evaluated by the local employers
1	Idea capture through own and third party perspectives and viewpoints depending on Explorative Phase, facilitated by intra-organisational means and external facilitators			
2	Idea capture depending on roles and business purpose Conduct of idea based business modeling Agreement upon context and relevance			
3	Comparison with enterprise strategy and values Comparison with initiation criteria and motifs for Smart Mobility			
4	Determining the transformation maturity Measuring maturity and deployability of the transformation capability based on the overall readiness for change and required skills and competences			
5	Deliverables: • Idea map • Agreement has been made about continuing or not			

Fig. 3.4 Smart Mobility Checklist for the Explorative Phase

Surveys, and also experiences from life in the city, quickly provide the current status of the city. This one has many facets.

The surveys and analyses result in a pretty clear picture about the status of the city: The demographics of the city have not changed fundamentally. However, businesses have migrated and work is undertaken locally less and less. In one of the neighbouring towns, for example, a production company has established itself. The growing concentration of specialist companies in the neighbouring town is not only attracting the interest of a training centre. The neighbouring village tries to find skilled workers who live in the area. A shuttle service not only transports the employees of the individual companies to and from work, but also cooperates with the local shops and thus provides additional services for the commuters in the workplace. Those residents who work in the big city are confronted with traffic jams and delays. Older, long-established citizens are confronted with the question of shopping opportunities. In Johannes' city, there are almost no shops anymore and the city has developed into a residential area. Those who would like to continue living there are increasingly looking for their offers on the internet. Patience is required if purchases are made in shops within a certain radius and if larger deliveries are involved. Deliveries from shops in the big city to the countryside have become rare. It requires an appropriate workload for the haulage companies and thus a bundling of deliveries in order to set off or optimise the route accordingly.

As far as they're concerned, it's out of town. The way to the doctor, whether general practitioner or dentist, is long. The public bus timetable does not allow much flexibility. The residents of the local retirement home depend on the staff to arrange appropriate office hours and take care of the logistics. There's basically no time for that. The staff are scarce and the budget is even tighter. The city has an old town centre, but it is becoming increasingly isolated. The only pharmacy on site has moved to a new development on the outskirts of the town.

The suburban situation is now completed with a look into the neighbouring villages. Johannes and the team also take a look back at the early years of urban development and the commercial and industrial settlements on site.

Since Johannes and his colleagues are in a start-up phase, they are inspired by their own developments, but also by innovative projects from outside. For this purpose, they conduct a survey of the situation of their own city in relation to other cities and which domestic stakeholders are involved there. Local appointments help to exchange information. It is also important to identify the influencers on site. In a neighbouring town, for example, a company training program me promotes the research focus of a nearby institute. In evening courses, a local service provider also offers information events on the research results in order to position itself as a mediator and future sales expert in good time. When the results are transferred into a network image, the first focal points of interaction become apparent. A first assessment of the strengths and weaknesses of the relationship between stakeholders is another result.

City xyz					
Population[3]	0–5000	5000–10000	10000–15000	15000–20000	20000–25000
Area	0–4000	4000–8000	8000–12000	12000–16000	16000–20000
Population density [in QM]	0–4000	4000–8000	8000–12000	12000–16000	16000–20000
Rail network [in KM]	0–100	100–200	200–300	300–400	400–500
Number of stations	0–100	100–200	200–300	300–400	400–500
Public bus fleet size	0–5000	5000–10000	10000–15000	15000–20000	20000–25000
Daily journeys[b] by train	0–2000	2000–4000	4000–6000	6000–8000	8000–10000
Daily journeys[b] by public bus	0–3000	3000–6000	6000–9000	9000–12000	12000–15000
Road network [in KM]	0–6000	6000–12000	12000–18000	18000–24000	24000–30000
Road traffic density	0–4	4–8	8–12	12–16	16–20
Number of vehicles[c]	0–1000	1000–2000	2000–3000	3000–4000	4000–5000
Share of private cars[d]	0–600	600–1200	1200–1800	1800–2400	2400–3000

Example classification[a] Population in '000,[b] Daily journeys in '000,[c] Number of motor vehicles and[d] Private cars in '000
Depending on the application, the scale must be adjusted accordingly, for example in trips in '00 instead of '000.

Fig. 3.5 Morphological box for comparing smart mobility in cities

Comparable data of cities and catchment areas with a similar profile can, for example, be easily transferred to a morphological box analogous to ◉ Fig. 3.5 and maintained there. Depending on the comparison space, the number spaces must be adjusted accordingly.

In terms of challenges and questions, Johannes' city is not far away from the concerns of other cities. The orientation work mentioned above helps to understand which project plans and efforts other cities have undertaken in a similar situation and which have been less considered.

3.3 Setting Priorities

Context—Context—Context

The reference to mobility is becoming more and more important in the analysis of the initial situation in Johannes' city. Mobility now determines everyday life and leisure, progress, our livelihood and the future of a city. From the point of view of a commuter or, for example, a physiotherapist, there are different points of reference for mobility. Differences can already be found in the term mobility itself, which is often perceived differently, and even in Johannes' project team the term is discussed in a strongly position-related way. A colleague therefore suggests the creation of persona profiles.

As described under Sect. 2.3, the result of a persona analysis provides information about motivation, wishes, expectations, and needs in the mobility environment. It is also helpful to include your own environment in the persona creation process and to check for completeness using the persona setcard. It is also helpful to go into one's own business on site or to exchange information with customers and business partners. At the end of this project work, the team has 15 persona setcards at their disposal.

One idea that is both exciting and challenging is the use of autonomous vehicles. Autonomous driving is currently being trialled in some places and regions in Europe—see Sect. 10.1. This is happening not only in cities, but, in fact, in rural areas especially, because a rural area can be considered a learning field with a dedicated field of application that could be built up. What are the benefits for citizens? The investigations of autonomous driving are far-reaching. They range from technological, socio-political, legislative to financial and acceptance-related issues.

The second obvious idea is to meet the mobility requirements and motion profiles equally. without increasing the number of predominant travel means, but instead divide and conquer through an optimised approach. Whilst meeting needs autonomously is technologically interesting, and sharing rather than owning is equally economically and ecologically interesting, focusing too much on one element could be strategically the wrong decision. The persona setcard is a good addition to the determination of the influencing factors, see Fig. 2.2, which can be found under Sect. 2.2 persona setcard that was introduced.

For projects like Johannes', the means of transport available in Johannes' city play a role. His team uses the template and creates an example from a colleague's point of view. This helps in the survey of citizens and business people in the city. The colleague evaluates the result for determining the factors influencing the choice of means of transport according to ◉ Fig. 3.6.

The "1-out-of-4" concept aims at evaluating the variables and additionally what acceptance of the mobility mix looks like. In addition to the question of suitable and available media, the focus is also on financing, legal requirements and the development of an adequate business and operating model. What a city like Johannes' needs, and can cope with, in terms of capacity is subject to our evaluation schema, in which we look into the factors influencing the choice of transport means from the point of view of a city without train connections ◉ Fig. 3.7.

Interestingly, the first assessment does not include rail when it comes to the mobility mix. Rural regions have different priorities, as an exemplary analysis shows ◉ Fig. 3.8.

The project team plans to better adapt to the traffic and service offers of transit passengers, i.e. business travellers and those who cannot find adequate charging stations and other sufficient services on the nearby motorway. They interview business partners of companies located in the neighbourhood. In the case of a company domiciled in the country, there are other influencing factors that count for their employees and business travellers, as shown in ◉ Fig. 3.9.

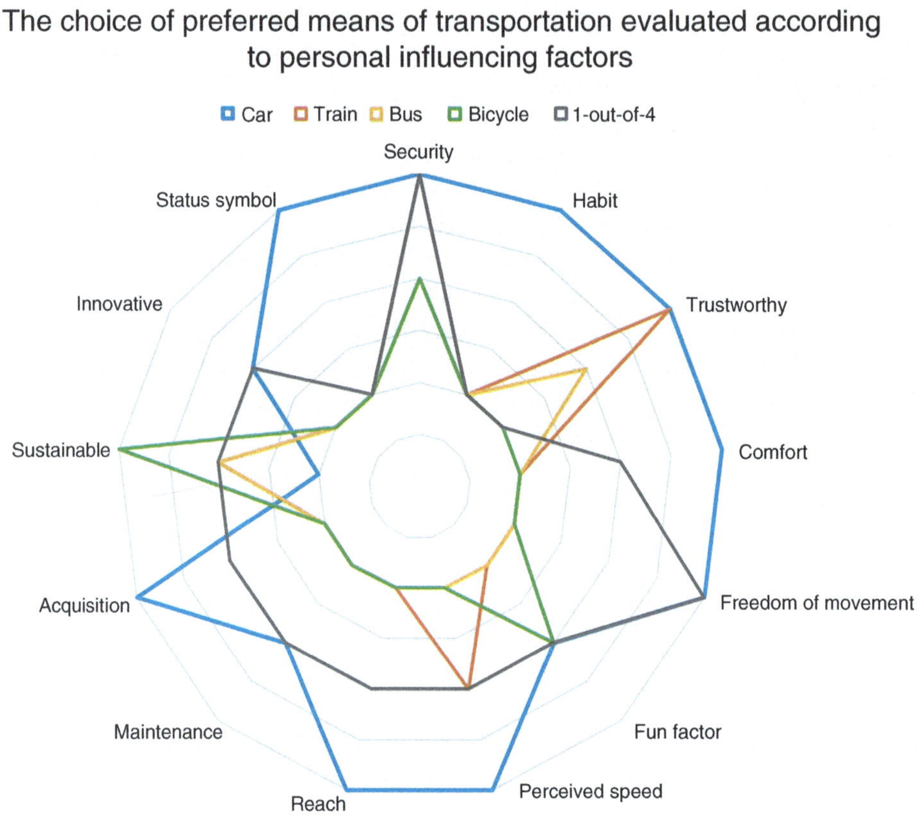

Fig. 3.6 Example for determining the factors influencing the choice of means of transport

The team is increasingly working on the MaaS concept. Is it not possible to combine overcapacity in the private car sector with requirements for services for transit passengers and short stays? A daring thought!

The effect formula of MaaS (cf. [3], p. 219) serves colleagues as a starting point for their considerations. It serves to verify the above consideration of considering a mobility mix of car, bus, and bicycle. The effect formula also sharpens the focus on mobility offers to previously unknown user groups.

In order to examine the adjusting screws of MaaS, part of the group concentrates on mobility consumers (◉ Table 3.2) and the second part on mobility providers. Where necessary, they note conspicuous features, local conditions or extend the characteristics.

From the mobility provider's point of view, the characteristics are outlined by disposition costs and cost optimisation (◉ Table 3.3).

It quickly turns out that the survey effort for the team is less on the consumer side. The selection and description of the personae does indeed involve a lot of effort and already arouses desire in the interviews with entrepreneurs, senior citizens and commuters.

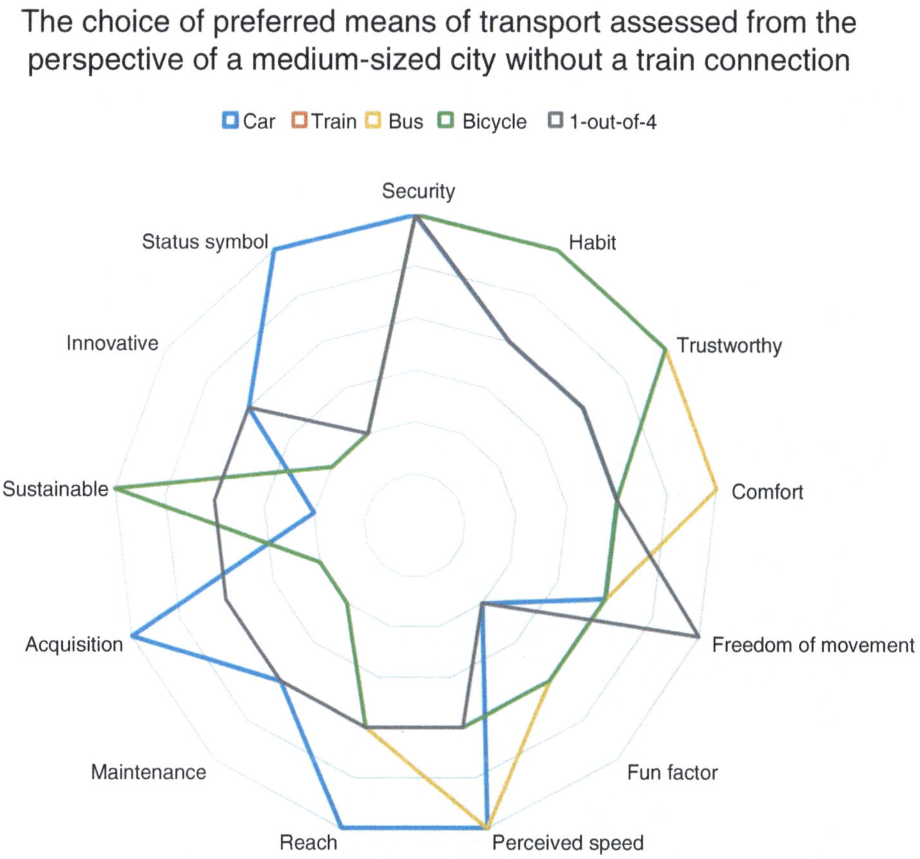

Fig. 3.7 Factors influencing the choice of means of transport from the point of view of a city without train connections

Nevertheless, the proximity to the location and the clarity of the design elements listed above from ◉ Tables 3.2 and 3.3 make it possible to conduct the interviews quickly and compare them well.

The colleagues did not reckon with the effort involved in recording the mobility providers. Here it is important not only to include the classic local and regional offers. It is also important to take account of movement patterns and demand patterns when shopping in other regions. The provider profiles examined also include the distances travelled by commuters and where they traditionally have stopping points, such as in company car parks or on shopping trips, before they make their way home. One of the annoyances for consumers, for example, is the long delivery times of a furniture store located in the region. Another annoyance is the lack of a shuttle service for trainees. While this is used in the neighbouring town, in the area surveyed trainees have to organise transport themselves, entirely independently of the companies or the associated vocational school. Couldn't the

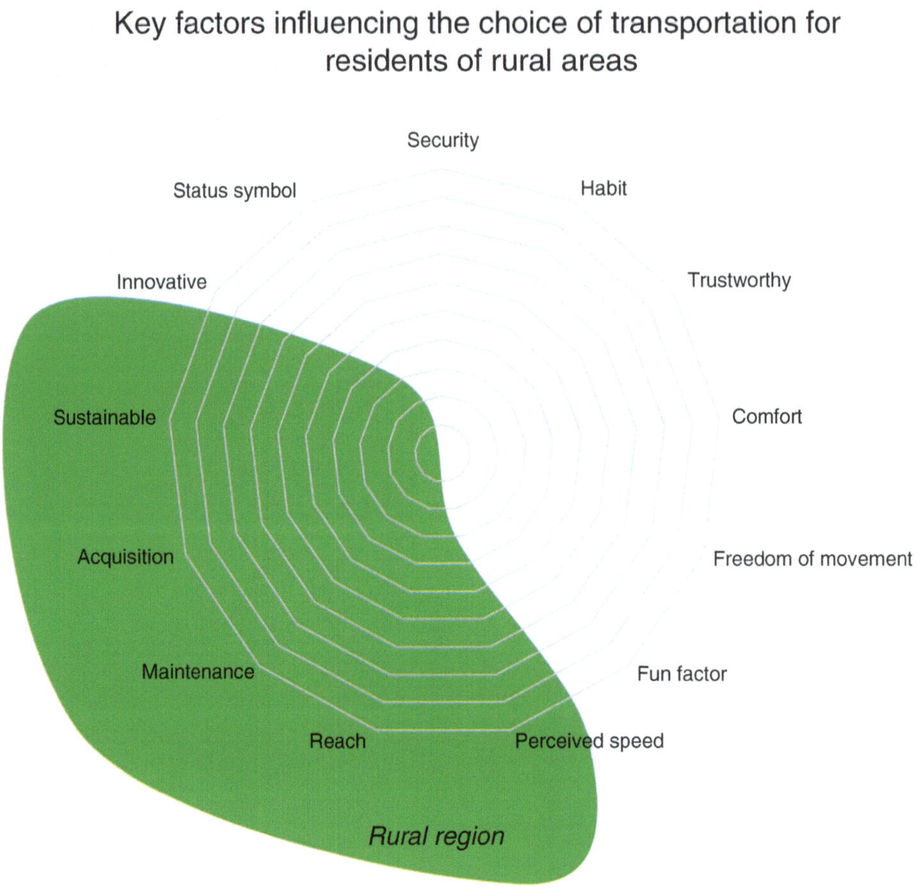

Fig. 3.8 Factors influencing the choice of transport mode in rural regions

company's own fleets also be offered in the event of missing vehicles or critical operating times?

The considerations mature during the survey phase: offers from small forwarders or rather their empty journeys and "remaining places" are also examined. For the latter target group, colleagues use their local knowledge and their own contacts—the market is diverse, and forwarders and dispatchers are constantly under pressure due to the tight timing of delivery cycles and the traffic situation in the agglomerations. A large number of data entry forms are quickly stacked on the project table and the timetable for the interviews testifies to the shifts that have resulted in part from a lack of availability or the search for suitable and willing contacts. Often, established entrepreneurs demand a justification for what the interviews are used for.

Johannes and his team set themselves no limits in the survey phase! Irrespective of the legal situation or political-strategic crosscuts, they first want to ensure transparency. Transparency in the sense of the possible. This gives you a holistic picture. They examine

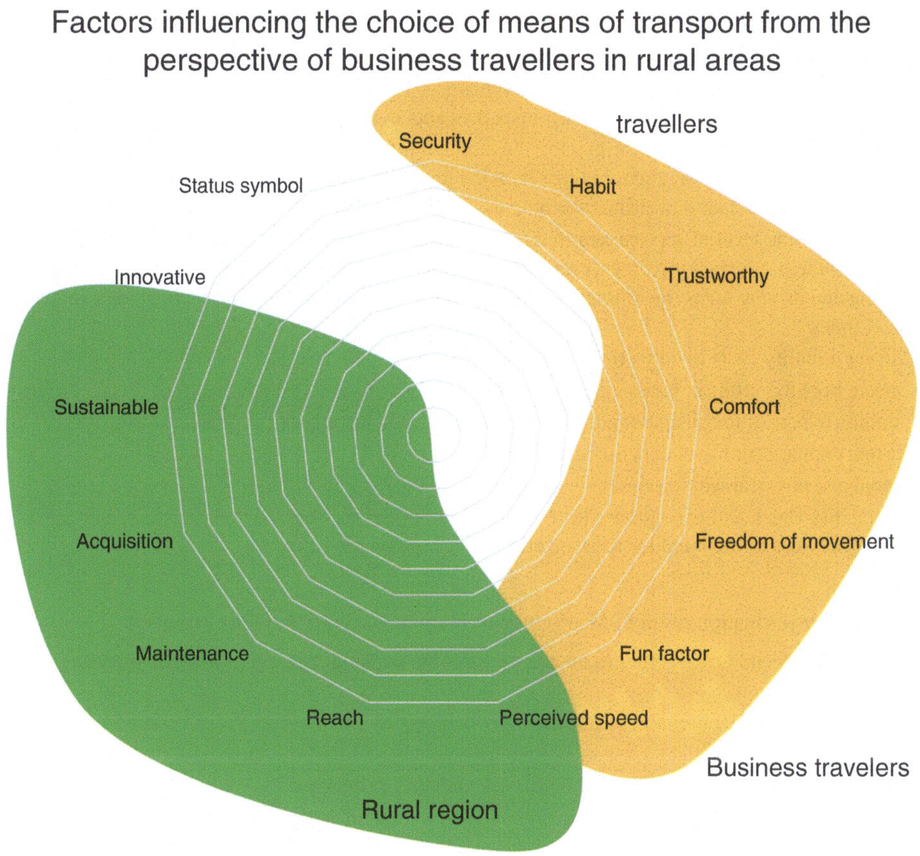

Fig. 3.9 Factors influencing the choice of means of transport from the point of view of business travellers in rural areas

not only personal needs but also freight needs, they examine mobility needs according to pathways and frequency, according to the bundling of needs and the efforts behind them. The naming of advantages and disadvantages of each means of transport is also part of this. They want to leave out a division into personal and freight needs for the time being. In the first interviews, a merger was evaluated as positive and beneficial. Too early a division into two could conceal optimisation and innovation effects.

They also hope to see a willingness to talk and an opening up of those who are interested in offers, such as from suppliers that are being pushed further and further to the edge. The most recent decision of the Karlsruhe Administrative Court in favour of the Bad Liebenzell [4] and the establishment of state-wide taxi associations in Switzerland [5] show the hardships from the point of view of the drivers. If the people of this town were to consider themselves, if we were to establish a new supplier, this would create further competition for the established suppliers on the spot and on the commuter routes, wouldn't it?

Table 3.2 Research result of MaaS implementation efforts for mobility consumers

Planned or completed routes to the region or work location
Actual or planned routes at the place of residence
Number of monthly and annual public transport tickets in the population
Definition of the service level for placing the demand for demand, in particular taking into account the time and money required for booking and with regard to punctuality and the use of relevant media (card, app, website, machine etc.)
Definition of the level of service required to meet requirements on site, in particular taking into account punctuality, waiting times, connection options, timetable coordination, cleanliness of the waiting and driving areas, recommendations to others, possibility of using tickets when changing travel times
Current mobility costs in euros from a persona perspective
Current booking effort in hours and euros from the point of view of a persona
Planning expenses for business, purchasing, sickness-related and holiday trips
Current expense for travel expense accounting
Virtual and physical waiting times
Availability check and expenditure for the use of additional services, such as a citizen call car or driving services of the local DRK association or of honorary drivers

Table 3.3 Investigation result of MaaS implementation efforts for mobility providers[a]

Disposition expenditure in order to be able to meet the demand regionally or to the pick-up or destination location
Purchase of receipts, purchase, or rental of payment systems and billing (card systems, etc.)
Fees and charges
Service level for positioning the availability of a means of transport
Transaction costs for bookings, such as manual and digital efforts for ticket booking and accounting
Waiting times due to timetable changes, illness, loss of means of transport, weather-related postponements
Sustainability balance of the company's own fleet or the means of transport of third parties used on its own behalf
Mobility balance of a route or a means of transport

[a] In addition to local public transport, local or regional taxi companies, providers of voluntary travel services and car services, the regional representation of a rental car company and the offers of parcel delivery companies and the fleet offers of companies and shops in the surrounding area or the nearest city are added

What does it look like in terms of sustainability? Measured against the distances of 30–50 km and a high volume of traffic, fuel consumption and emissions can be quickly estimated. For example, Swiss railway company, SBB, shows a 100 km return rail route with a saving of 5 litres compared with an alternative car. The figures may vary depending on driving style, weather conditions and traffic jams, but they still speak for themselves. Carsharing e.V. conducts its own surveys for this purpose, but also draws on data from Great Britain [6].

To find out what's going on, Johannes and his colleagues have already collected data from the Federal Environment Agencies for advice. They would still like to carry out their own measurements. A holistic approach includes the maintenance costs of private, shared and public transport, their utilisation maturity, maintenance and deployment costs, and their energy pass. However, the sustainability of urban and regional traffic volumes measured against the polluters and the fulfilment of mobility needs also aims to demonstrate the benefits of alternative mobility options to local citizens.

3.4 AIM—Helping Mayor Johannes

Marco Linpinsel and Pascal Luder

Marco Linpinsel and Pascal Luder learned from Johannes about his efforts. After engaging with Barbara Flügge at the Smart Mobility Strategies and Management master studies at ZHAW Winterthur, Barbara asked Marco and Pascal to look into a sustainability measurement service for Johannes and his city.

In the nineteenth century, the Englishman E. Howard developed a green urban planning concept in response to increasing industrialisation—the "garden city". This concept was also implemented in Winterthur in the early 1920s and still characterises the cityscape today. Most of the small workers' houses were surrounded by small gardens, and today they are coveted living spaces. The spacious gardens of the high society villas also contribute to the abundance of greenery, and some are now publicly accessible green spaces [7, 8].

The Smart City of Winterthur is defined as "[...] an innovative, progressive, and connected city that puts people and the environment at the centre. It is characterised by a high quality of life and efficient use of resources. This is achieved by intelligently networking infrastructures with modern technologies and by involving relevant stakeholders. In doing so, social, economic, ecological, and cultural requirements are taken into account" [9].

To understand people's perception of the meaning of intelligent mobility, we conducted a survey with people who live and/or work in Winterthur and asked them what they imagine by the term "intelligent mobility". The results show that people imagine not only the general intelligent choice for travelling from A to B but, more importantly, the consideration of their own criteria and needs by putting their experience at the centre of transport decision—making, using data and technology. Important factors which should play a role in this process are time efficiency and the inclusion of all available transportation means, such as Public Transport, Scooter, Uber, Mobility, or their own car.

The feedback showed that the combined use of different means of transportation—which is described as intermodal mobility—is an important aspect of intelligent mobility, not only for the respondents, but also to solve challenges in the whole mobility sector. The objective of intermodal traffic management is to cover all mobility needs with the least additional traffic modes possible. Typical intermodal travel sequences are:

- Walk to the bus stop,
- Take the bus to the train station,
- Take the train to the next train station near the concert hall,
- Take a taxi to the concert hall.

With the increasing number of available mobility providers and their systems in Switzerland, such as Uber, Mobility (cars), SBB, ZVV (public transport), Bond, PubliBike, Carvelo2go (bicycles), Bird, Circ, Lime, Tier, and Voi (e-scooter), combined with the individual needs of the transport user, the planning of intermodal mobility is a complex task [10–12]. Figure 1 [13] describes the different aspects of intermodal and intelligent mobility and their interactions.

The process starts with a mobility consumption need: Peter needs to travel from Winterthur main station to the Fotomuseum in Winterthur. Since the museum opens at 11:00 on Sundays, the travel schedule needs to fit accordingly. Based on his personal needs and preferences, Peter's priorities for the journey are time, comfort, and cost, while travelling sustainably, which can be indicated in the app. The available transport modes for this trip are walking, taking a bicycle, taking a Lime scooter, taking a bus (lines 1, 2, 3, 5, 7, or 10), using the Swiss mobility rental car, using Uber, and/or taking a Carvelo2go [14, 15].

The "Digital Concierge", an intermodal mobility intelligence assistant, suggests feasible combinations based on the previous stated points. Peter's options are:

1. 1-8-2: 1 min walk to bus line 2, 8 min bus ride and 2 min walking.
2. 5-6: 5-min walk to the next Lime scooter and 6 min ride with the Lime scooter.
3. 19-min walk.

However, bus line 2 is out of service due to an accident. In the case of this incident, the Digital Concierge sends instant messages and adjusts the route. Peter can pick an alternative to get to his destination and adapt the preferences in his profile, if required.

There is a wide range of existing intermodal route planning and other smart mobility solutions on the market. Since those solutions have different characteristics, cover different aspects, and are available in different areas, this paragraph gives an overview to identify the best practice solution, and the potential for improvement and combination of technologies.

The Figure 2 [13] shows existing mobility providers and their characteristics in Winterthur and puts them into perspective in terms of availability in time and spatial availability [16].

Besides the above-mentioned mobility providers, there are also several mobility service providers on the market which offer intermodal trip planning to support the user with planning and comparing combined-mode journeys. Since only some of them operate in

Winterthur, Figure 3 [13] shows existing solutions from other locations as well. The key characteristics of an intermodal journey planner covers non-local providers. A key criteria list covers the following characteristics:

In the persona profiling study, we distinguished five personas. The following appeared in this research:

- Children & Teenagers
- Young Adults
- Adults
- Retired
- Silver Generation

A smart mobility solution only serves intelligently and becomes successful when customer needs are met, and the user is put at the centre of the experience. Therefore, an intelligent mobility assistant should be able to adapt to the customer's needs and offer services accordingly. A "one-size-fits-all" solution should not be the objective, as it would compromise on many existing user needs.

The previous paragraphs show that there is already a wide range of mobility providers and some intermodal journey planning solutions on the market. The solutions show different characteristics and levels of maturity.

One main issue is, on one hand, the lack of integration and interoperability between different providers, and, on the other hand, missing user centricity [17]. Besides the intermodal planning, the respondents of the conducted survey also mentioned the need for additional services, such as sustainable travelling, ticketing, and seat reservation. Some respondents also questioned whether they would use such an app at all, taking the already available solutions into account.

The solution we propose is "AIM". The AIM App stands for a user-centred Assistant for Intelligent Mobility in Winterthur. The characteristics of AIM, its functionality and different extension stages can be described as follow and is visualised in Figures 4a and 4b [13]:

1. Intermodal mobility planner:

Integration of all available mobility providers in Winterthur, including traffic, parking, and e-charging information, into one app. An easy-to-handle user interface, combined with detailed filter setting is important to satisfy user needs and convenience. Depending on personal preferences, the options are sorted by sustainability, time efficiency, price, number of changes, or number of entertainment options on the route. Options that do not match preference, for example because of a lack of accessibility, are not shown.

2. Integration of additional services:
   (a) Booking of additional services, such as seat reservations -> integration with SBB and ZVV.
   (b) Carpooling platform to share rides and socialise with other people from Winterthur -> integration with "TaxiTo" and "BlaBlaCar".
   (c) Recreational activity suggestions (concerts, exhibitions, fairs, museums, restaurants, bars, clubs, etc.) en route in and around Winterthur, based on personal preferences. Bundle prices with event and mobility ticket.
   (d) "A-Day-Away" suggestions: The location, maximum journey duration and means of transportation can be specified and a coloured map shows where you could go. Can maybe combined with 3c.) and integrated with https://app.traveltime.com.
3. Integration of payment options with different alternatives:
   (a) Pay-per-use model with "FAIRTIQ" integration for public transport and links to external mobility provider system. This option would be easier to implement but leads to an average customer experience.
   (b) Fully integrated ticketing in one app without necessary links to other applications. This could be done with either pay-per-use model, like "FAIRTIQ", but for all transport modes, or with a monthly subscription and bundle for the different means of transportation, similar to "Yumuv" or "Citymapper" in London. This option would be more complex to implement, due to barriers between different providers, but would be more appealing and customer-centric.

A visual outline of AIM is highlighted in Figures 4a and 4b [13].

What are the key elements to carry out a sustainability impact assessment?

Measuring the impact on sustainability is based on the following dimensions, namely economy, society, and environment.

In a key next step, each of the indicators were mapped onto the key characteristics of the AIM application—we recommend using visual mapping through the use of distinct symbols. Consider the following: an equal sign means no relevant influence, a plus sign represents a minor impact, two plus signs represent a medium influence, and three plus signs stand for a major influence. For this purpose, the possible changes compared to the status quo were estimated. The entire study is available upon request by contacting the chief editor of this book.

The increased use of multimodal trips, such as those provided by AIM, will certainly lead to lower greenhouse gas emissions. There are many reasons for this: shared usage of cars through carpooling or car sharing, less car driving, and instead an increased use of public transport or muscle-powered vehicles. Less car usage also reduces air pollutant emissions and noise pollution along roads.

The mobility needs of Winterthur residents are likely to be very well met by AIM, since the app works on a very individual level and a unified payment system would be a great improvement in user-friendliness. However, it is of fundamental importance that all relevant stakeholders are involved in the development and implementation as early as possible so that the app is accepted broadly. The simple and intuitive operation and the involvement of key stakeholders are of great added value for the city and its residents [18].

Cities like Winterthur and their mobility systems face various challenges, such as increasing levels of urbanisation, climate change, transport systems reaching their capacity limits, and a high variety of mobility needs. Due to the diverse range of inhabitants, many differing mobility needs must be considered. Needs profiles depend on the age, occupation, health, interests, personal interests, hobbies, and habits, of the individual, to name a few. Intermodality, sustainability, health and electromobility will certainly play important roles among those needs.

In recent years, the public transport offers constantly expanded with increasing SBB passenger frequencies and additional mobility service providers, such as E-scooters, E-bikes, cargo bicycles, and carpooling solutions entering the market. These solutions are characterised by a variety of different features and use cases. Intermodal planning solutions, like Google Maps, Citymapper, and local solutions like ZüriMobil, try to combine these versatile mobility solutions to improve the mobility experiences and tackle the mentioned challenges.

But, overall challenges, such as capacity limits and climate change, can only be overcome if mobility solutions are seamlessly integrated, with their different characteristics combined in a way that user acceptance is increased, and personal needs are met. So far, existing intermodal planners seem to lack this seamless integration and user centricity.

However, it is still unclear whether an additional or rather regional solution will lead to more acceptance and usage of such a solution and result in a shift towards more sustainable mobility, even if it might be of interest for the city of Winterthur from a marketing perspective. Therefore, conducting thorough market research should be considered, and existing mobility service providers could be approached in order to use, integrate, and adapt their solutions to achieve the set objectives.

3.5 Taking Off

After a citizen survey and a review of all analysis results, the decision is made to establish the mobility mix driven by our 1-out-of-3, excluding public railway services ◉ Fig. 3.7.

The next step will be creating an organisational overview of who is eligible for which tasks according to the MaaS concept. The so-called effect formula of and the implementation of a 2-day business model workshop have established initial assumptions for the operational and financial expenses. The operational characteristics will be tested in a pilot project. This includes determining the range of vehicles, and the willingness to use a smartphone app and to engage in vehicle sharing. Especially for the financial and operational tasks, the following is considered to involve external experts and innovators in a targeted manner. The list of tasks for the first test phase is impressive:

- Implementation of a Smart Hacks with students and trainees of the specialist companies and universities in the vicinity of the city
- Call for tenders for a crowdfunding project to test e-bikes for senior citizens and cargo bikes
- Convince the representatives of the public transportation companies who are responsible for a. bus operations, to use the space for transporting purchases and parcels, and b. bus schedules, to share the bus timetable using an interface to digitally integrate the bus timetable into the MaaS app.
- Convince car owners and taxi drivers to take part in the test phase and to adopt their own accounting model in the sense of 1-out-of-3.
- Conducting the Business Model Innovation Workshop mentioned above to prepare a blueprint for financial and capacity expenditure and obtain relevant approval for testing.
- Planning of the pilot phase with regard to duration and test scope
- Presentation of the project results and the project plan at the citizens' meeting

The aim is to make optimum and full use of existing capacities, as this was not previously the case. Offering the population flexible, more comfortable, and yet cheaper, mobility is a declared success criterion.

Are the Test Phase and the Main Tasks Described Sufficiently?
Consideration is being given to testing new charging technologies that enable mobile charging. To what extent, for example, will car owners of electric cars agree to make their vehicles available for car sharing? The scheme would need the appropriate marketing and the announcement of the project in social networks. If the city decides to test new charging stations, it will require location, financing, and a test clientele that are located far beyond regional borders or require fast charging functions for distances of 200 km and more. The team meets with a winegrower hotel nearby to get an impression of the professional groups and how the hotel guests arrive. The project colleague, who runs a bakery, has also been thinking for some time about taking over the disused butcher's shop next to his business and offering a café as well as a pick-up service for Sunday and holiday deliveries. There'd be plenty of room.

The team deals centrally with the question of whether the inhibition thresholds identified at the outset, which includes the lack of local service providers, as well as the lack of attractiveness for investors and foreign settlements, have not lost their significance with the focus on the test phase. Additionally, they consider whether one has concentrated too much on the mobility factor.

Since further volunteers have been found through the calls for the City Hack and the involvement of training companies, Johannes and his colleagues decide to take a closer look at the history of their city. How did the city develop and grow? Which professions and companies were on-site at the beginning and which ones migrated when? Much of the urban heritage has been neglected or forgotten over time. Crafts enterprises had not invested in new tools and processes in a long time. Instead of joining forces with other companies to form cooperatives, attempts were made to ensure their survival by isolating them. The lack of willingness and also the lack of successors led to emigration and closure. The innovative power that had contributed to the growth of the city at the time should be brought back to the surface.

We can be curious about the results of the first test phase. And how a first charging station for emobility will impose innovative strength on the city. If, for example, a mobility lounge is set up for inbound and outbound business travellers, as well as for business travellers in transit but currently waiting, the community gathering gains a whole new meaning. What would it be like to combine the mobility lounge with an innovation lounge, where companies present their innovations and create an environment in which any trainee's idea can be presented to the travellers in an attractive ambience? Serving for mobility would gain a whole new meaning!

3.6 DoktorB's Checklist

Important Basic Considerations

- What does the target image of your city look like?
- What advantages does your city offer to an investor and what to the population?
- The 5-point plan offers a good basis: with which focus do you plan to start?
- Do you know your customers?
- How easy is it for your customers and visitors to reach the city?
- What recommendations do you have for Johannes and his team?
- What projects are you currently tackling with your fellow citizens?
- Do you see your location as part of a mobility network?
- In your opinion, which persons should be examined by Johannes and his team?
- What does your own location profile look like? Do you already have a persona setcard?
- Would AIM be of interest to help your mobility measures become more sustainable?

References

1. Federal Ministry of Transport and Digital Infrastructure – German-Polish Spatial Planning Portal (2018) Strengthening and deepening the links. https://www.kooperation-ohne-grenzen.de/de/staerkung-der-verflechtungen/. Accessed 17 Dec 2017
2. W.I.R.E Web for Interdisciplinary Research & Expertise (2010) DOMINO. Handbook for a sustainable world. https://www.thewire.ch/de/publications/domino. Accessed 17 Dec 2017
3. Flügge B (ed) (2016) Smart mobility – trends, concepts, best practices for intelligent mobility. Springer Vieweg, Wiesbaden

4. Black Forest Messenger (2017). https://www.schwarzwaelder-bote.de/inhalt.bad-liebenzell-buerger-rufauto-gericht-schraenkt-betrieb-ein.50478d18-bbe1-4fb3-89f6-ad8663072f8e.html. Accessed 2 Nov 2017
5. NZZ Website (2017). https://www.nzz.ch/wirtschaft/uber-konkurrenz-ld.141764. Accessed 25 Jan 2017
6. Bundesverband Carsharing bcs (2017) Carsharing relieves the burden on the environment and transport. https://carsharing.de/alles-ueber-carsharing/umweltbilanz/carsharing-entlastet-umwelt-verkehr. Accessed 16 Dec 2017
7. Website Garten.ch. https://www.garten.ch/fuehrer-durch-winterthurs-gaerten/quer-durch-gartenstadt-winterthur. Accessed 12 May 2022
8. Stadt Winterthur (2022) Stadt Winterthur. https://stadt.winterthur.ch/themen/leben-in-winterthur/planen-und-bauen/staedtebau-und-freiraum/gartenstadt/ein-leitfaden-fuer-die-gartenstadt. Accessed 28 May 2022
9. Stadt Winterthur (2018) Strategie smart city Winterthur
10. Fluegge B (2017) Smart mobility – connecting everyone. Springer Vieweg, Wiesbaden
11. Website comparis.ch. https://www.comparis.ch/carfinder. Accessed 17 May 2022
12. Website /elektromobilitaet/e-scooter-sharing. Accessed 17 May 2022
13. AIM (2023). Outline of AIM figures. https://www.barbara-fluegge.com/Mobilize4Data. Accessed 12 June 2023
14. Website Forum Bikesharing Schweiz., https://www.bikesharing.ch/de/bikesharing/interaktive-karte. Accessed 19 May 2022
15. Google Maps (2022). https://www.google.de/maps/dir/Winterthur,+Hauptbahnhof,+8400+Winterthur/Fotomuseum+Winterthur,+Grüzenstrasse,+Winterthur/@47.4987838,8.7272307,16z/data=!3m2!4b1!5s0x479a9990a8703ee5:0xaed54251a6c9fd31!4m14!4m13!1m5!1m1!1s0x479a997779effff9:0x66c599d28504. Accessed 17 May 2022
16. Eidgenössisches Departement für Umwelt, Verkehr, Energie und Kommunikation Bundesamt für Verkehr (2018). Literaturrecherche Handlungsempfehlungen «Mobility as a Service (MaaS)». AWK Group, Bern
17. Nemtanu F, Schlingensiepen J, Buretea D, Iordache V (2016) Mobility as a service in smart cities. In: Responsible entrepreneurship vision, development, and ethics. Bucharest
18. Carabias-Hütter PV (2022) Interviewee, Mobilität und Smart City Winterthur. Conducted 13(6):2022

Mobile Adventure Rooms

4

DoktorB Barbara Flügge and Hans Roth

Abstract

Where do our experiences take place today? How do we share them today and where could we share them tomorrow? Consider the car as an experience space, as a mobile chat room, showroom and the new supermarket: these and other associations are not only occupied by consumer goods and product designers. The automotive industry, which to date has measured itself by the number of horsepower units, the diversity of model series and a 1:1 addition to a sedan series with an SUV model, suddenly finds itself in competition with information and entertainment providers, as well as mobility providers, when it comes to the favour of the mobility consumer. Will the OEM disappear from the windshield, and will the smartphone provider instead drive our mobility appetite? We go to a café and quickly searched the web for a new coat. Ten minutes later a car picks us up and drives us purposefully to the top 3 boutiques in the city to try on the coat in the appropriate size and colour that we have uncovered in the meantime. A utopia? Come and let us entertain you.

D. B. Flügge (✉)
digital value creators (DVC), Bottighofen, Switzerland
e-mail: barbara.fluegge@dvcconsult.com

H. Roth
Harman International Automotive Division, Karlsbad, Germany
e-mail: hans.roth@harman.com

B. Flügge (ed.), *Smart Mobility in Practice: Good Car, Bad Car, No Car – Is this the End of Nurturing Our Mobility DNA?*,
https://doi.org/10.1007/978-3-658-43171-6_4

4.1 Experience and Linger

Where do our experiences take place today? If one asks advertising experts, the perception of an experience begins with the digital and physical placement of an idea for a product through videos and images, explanations, the recording of product-related smells and music features and much more. If we are in a supermarket, we can certainly confirm the variety of perceptions and the incentives such as advertising, discounts, videos, etc. that lead us to go there.

What happens when we're on our way? If we are guided by our preference profile that includes advertising next to personal interests and behaviour-driven suggestions, a sophisticated system will strengthen and control our perception. As soon as we get close to a department store, our smartphone plays holiday music and shows us fake holiday photos of the beach, the Eiffel Tower or the Alps. With digital comparison, the holiday mood targeted by such advertising, and the necessary desire/need for a suitable wardrobe, DoktorB's own role model web shop, HolidayChic, carries out a wardrobe check at our home:

- What is available in our holiday wardrobe, what is suitable, and what, on the other hand, will never find entrance into your suitcase for the holiday?
- What meets your preferences and what is appropriate? Each piece is neatly registered in a digital format: smart tags, aka intelligent markers, are attached to the clothing and managed on the platform.

HolidayChic determines the wardrobe gap, advises us as to which pieces are missing, acquires them and delivers them to us. This leaves only actually packing and transporting our suitcase to the airport to manage on our own. All preparatory steps including decision-making, selection, purchasing, delivery and payment take place digitally. Analysts and information service providers collaborate to serve us best. Let's spin the scenario further. In the future, we might digitally select the suitcase, purpose, and occasion. We then receive the recommendations of what to pack and how. Once finished, the suitcase is delivered to the point of destination, without stress and without lugging it around in our hotel room at the holiday destination. And at the end of the holiday we fly back home relaxed with only travel souvenirs in our luggage and welcome back a 360° service experience.

How far-reaching is our desire for experience? Are To Dos, such as searching, experimenting, and choosing, even part of having an experience?

When the Anthropology boutique [1] opened its doors, visitors were surprised and pleased because of the hitherto unforeseen, surprising ambience they encountered. The product mix of clothing, accessories, furniture, gifts, and decorative articles was arranged in a carefully thought-out manner according to colours and moods. Anthropology not only attracted the attention of customers and the curious, but also made itself heard by marketing experts.

Virtually, Anthropology gained a rush of enthusiasm due to an elaborately prepared and high-quality digital, as well as printed, available catalogue. The presentation was based on the narrative and representational style of picture books. This was accompanied by a narrative plot that led the reader to places of longing with product and decoration recommendations tailored to them. Styling slots were offered. These decision aids helped those who didn't know how to decide and didn't want to reveal their own taste. The styling appointments were also a welcome occasion to meet like-minded people. Clever content marketing, blogs, event announcements and pinterest markers project the Anthropology experience into words and images that the ideal anthropology fans wish for.

4.2 Designing Experience Spaces in the Car

How do we create mobile experiences—not at Disneyland, nor European fun parks, nor at LEGOland, but in the car, in our car? What role does the car play alongside simply to becoming the space provider? With regard to the physical placement of shopping experiences, it is particularly worth taking a look inside the vehicle once potential buyers are addressed directly on the basis of their preferences, habits and the appropriate ambience. The ambience is either created by the car itself or added to the car as a virtual and physical cover. The ambiance result draws on existing experiences from static spaces such as Anthropologie's shops.

What does a mobile car experience look like? I am interested in the views of HARMAN. Let's look back to the International Automobile Exhibition 2017 and the Electronics Fair in Las Vegas 2018. I met Hans Roth from HARMAN [2] in Frankfort, Germany.

DoktorB Barbara Flügge: I am delighted that you are joining me on a journey into the world of cars. How is in-car digitisation thriving given HARMAN's market strategy?

> Hans Roth: Sidney HARMAN founded HARMAN Kardon in 1953 and the following year launched the world's first hi-fi receiver under this name. The first automotive products followed in 1966. The company has been called HARMAN International since 1969. Since March 2017, HARMAN has been a wholly owned subsidiary of Samsung Electronics. Well-known brands offered under the HARMAN umbrella include – in addition to HARMAN Kardon – JBL, Infinity, Lexicon, and Bang & Olufsen Automotive. After the takeover of Becker Autoradio in 1995, the group developed more and more into a provider of navigation and infotainment systems for the automotive and mobility industries. Through the integration of HARMAN, Samsung gains direct access to the leading automobile manufacturers. Today, more than 50 million vehicles are equipped with HARMAN audio and infotainment systems.

Barbara Flügge: Let's take a look at the other core areas of HARMAN, Mr. Roth. I noticed that the role played by HARMAN to date has changed considerably.

Hans Roth: Our automotive fields of activity have expanded considerably in recent years. The primary goal is to create a holistic user experience. For this we merge these elements.

Barbara Flügge: These elements are as I have studied them prior to our interview:

- The perfectly integrated digital cockpit
- Connectivity and Telematics
- Cloud Platform and Personal Assistant
- Cyber security
- Driver assistance systems and autonomous driving
- Car technologies for the perfect sound experience [3]
- Soundsystems with the focus to offer users even more listening experiences.

Hans Roth: Yes, that's our solution portfolio.

Barbara Flügge: Autonomous driving sounds exciting—will you build your own autonomous vehicles?

Hans Roth: No, the production of our own vehicles is not the focus today. Nevertheless, as the Samsung Company, we take a close look at what the big players like Google and Apple do. Their research into new business areas in the mobility segment is characterised by the manufacture and sale of autonomous vehicles.

Barbara Flügge: Where do you see further trends?

Hans Roth: Well, on the one hand the content moves to the cloud and the personal assistant is also in the foreground. ([4], p. 101 f.)

Barbara Flügge: By accessing content at any time and in any mobile or static state, we experience the car in a different role. What's so interesting about this new role?

Hans Roth: Well, regardless of whether we're talking about a semi-automatic or autonomous car, the car is now presenting itself as our new home. The car features and functions that we admired at the Geneva Motor Show, the IAA or the Chicago Motor Show, could be more likely to be found in living rooms than in passenger compartments.

DoktorB's Conclusion

Our questions in everyday life dissolve from the question of locomotion. Instead, we focus on questions such as "What can I do at home?", "Who can I communicate with in the office?", "What services can I consume?", and "How can I increase productivity?". More and more attention is being paid to cooperation. Early offers such as the one from Waze

[5] shows us how we can share content on the spot. At the beginning, the sharing focus was on traffic and congestion content, recommendations for alternative routes or indications of possible danger zones. In the future, we will live out the urge to share and consume much more through film and music services and social exchange. The latter can take place within a vehicle, across several vehicles or between bypassers, friends, and strangers. The way of sharing and communicating, as we know it from social networks, is finding its way into our means of transport!

Our workplace becomes even more mobile: The car is gaining in importance because we equip it with mechanisms that enable us to gain time and energy for other activities. The quasi-omnipresent access to everything we need takes place through providers. The latter can be found in the automotive and service industries, in the consumer goods and tourism sectors, as well as in offers by individuals and influencers. Influencers, their social network and their status in the social networks, play a significant role in the acceptance and rejection of new offers.

With regard to entertainment, more and more is streamed, leased, used, shared, or offered on-demand. What does this mean in concrete terms? Passengers in a car or the driver himself can download, forward, share, or accept offers, for example, to trade their data for a discount or for add-on services. Analogous to the innovations and changes that we observed in the music and film industries, technical progress is an elementary building block for the digital transformation of the automobile: 10 years ago, the design of so-called digital rights management was a matter of concern, whereas 6–8 years ago, the size of hard disks in the vehicle was of decisive importance for providers of entertainment functions, such as the ones that HARMAN offers.

4.3 The Ultimate User Experience

We proceeded with the interview:

I would like to know what the car of the future, aka future car, looks like in 5–7 years?

> Hans Roth: "The customer who today enthusiastically watches YouTube videos for the handling of the new drilling machine and shares product recommendations with his online community is now looking for digital feasibility in the car as well."

Barbara Flügge: Everything that I can do technically, I want to do in the car as well as with the smartphone or on the desktop. This means that we increasingly expect a simple, intelligent, and forward-looking operability of functionality in the vehicle.

> Hans Roth: "And instead of our smartphone, the car becomes our new touchpoint. It not only offers us larger displays than tablets. All we need to do is take a look at concept cars like the OASIS car from Rinspeed". [6]

Barbara Flügge: If the goal is the maximum user experience, will HARMAN be more than a provider of sound systems? Mr Roth, you are talking about the "modes" of HARMAN. Are you referring to modes as becoming the immersive opportunities that arise through technologies for distinct applications?

> Hans Roth: "Correct! The "HARMAN modes" range from Drive Mode, Entertainment Mode, Assistant Mode and Productivity Mode to Social Mode. The latter brings the webshop into our car, skilfully shows us the price comparisons of our favourite coat with pictures and background music, and asks our best friends in New York and Zurich for their recommendation. Likewise, the information systems available on the market show us the range that the networking of location, preference and mobility behaviour can have.
>
> Another mode is added with level 5 of autonomous driving: the Relaxation Mode. From the yoga exercise in the car, the 5-minute meditation for the freight expert, formerly a truck driver, everything is conceivable. The control of the interior ambience puts us in the right mood."

Barbara Flügge: Why, Hans, does the car as a means play such an elementary role in Relax Mode?

> Hans Roth: "Well, the possibilities of interaction and display presentation cannot be achieved with the smartphone, nor can vehicle-specific functions, such as seat position, air conditioning, and light mode, be steered without the sensing for relaxation. Large displays allow our worlds of experience to be conveyed much more skilfully."

Barbara Flügge: The other passengers are offered a user experience that no longer depends on whether I am travelling, nor what car brand I am travelling with. This also leads to the distancing from OEM brands mentioned at the beginning. We are moving closer to experience brands. Will those OEMs that meet our taste and fulfil our requirements be the winners, especially if they succeed in offering a tailor-made experience to us?

> Hans Roth: "Exactly! The differentiation of the OEMs no longer takes place according to the question 'which car, which performance', but 'who delivers the best experiences?'."

Barbara Flügge: Let us include the end customers, such as drivers and accompanying drivers, in the discussion from HARMAN's point of view. Where does customer loyalty begin for HARMAN?

> Hans Roth:"Let's think of a typical daily routine: In the future our personal assistant will take us from home to the car! When we go where, which dates we have and which outfit is best for it, assistance systems like Alexa [7] take care of that."

Barbara Flügge: That also means, Mr Roth, that, as in our prototyping and example of the digital concierge, we relate the preferences of mobility users to the context, so that we develop and offer a service portfolio? That would also mean that HARMAN would be at the forefront of end customer decision-making and would be looking to collaborate with Alexa and other digital assistants?

Hans Roth: "Exactly!"

Barbara Flügge: I would like to talk about the so-called mood controller which is serving us dependent on our mood.

Hans Roth: "Now mood controlling, Ms Flügge, and control are essential in order to serve the consumer for the entire mobility experience. Therefore, we need to collect all context-relevant characteristics such as preferences, the mood as mentioned above, the personal situation and geographical position. With regard to control, the stocking of all relevant characteristics is omnipresent. This 'always knowing everything' does not only happen with us, but also takes place in the environment. So, I can have one mood transferred from me to another."

Barbara Flügge: Let us take the following example: I am on my way from St. Gallen to Zurich. According to the calendar, I should actually call someone and clarify a preparation appointment. My assistant in the car, the digital concierge, checks the situation I find myself in with the vehicle beforehand. Since a dangerous situation arises—construction site with a narrowing from two to one lane—my assistant decides to leave me alone. A postponement of the appointment is suggested to my interlocutor.

Hans Roth: The assistant and the car become my confidants and act as if they had been working with me for a long time. A similar picture emerges when we look at an executive and his executive assistant – through long-standing cooperation, the assistant knows what is opportune when and how and where to adapt everyday life. The big difference, however, is that our digital assistant accompanies us – through and along every situation and constellation.

Barbara Flügge: This requires a so-called individual mobility account or assistant account. Let's move on to another aspect. We have been exchanging about brand preference, the perception of our taste and how we deal with external developments and progress. You had an interesting example of this!

Hans Roth: "Let's look at Audi's example in connection with the type-dependent lighting control system that was launched on the market years ago. The Unique Selling Proposition (USP) is not determined by the driver. The new lighting control system is perceived outside the vehicle. It is visible to the environment, other drivers, vehicles, or pedestrians. Nevertheless, as an Audi buyer, I'm prepared to pay extra for it. As a driver, I share the experience without seeing it for myself. Audi has thus created a distinguishing feature that has revolutionized the industry.

Audi has thus also created emotionality for me in the Audi vehicle: I am aware of the anticipated positive perception of those (in the streets, on the pavements, etc.) who recognize the innovation in lighting control from the outside and thus identify me as the driver with such progress. I see a vehicle with the same one and I want it."

Barbara Flügge: But this would also mean, Mr. Roth, that emotionality will become the differentiating feature of the future, so to speak "advancing through emotionality"—compared to Audi's marketing claim "advancing through technology"?

> Hans Roth: "Exactly! Above all, it also means that competition has to adapt, indeed has to adapt and thus catch up again. Herein synergetic effects and mutual inspiration and new forms of innovation emerge."

4.4 Role Models of the Future

Let us now look at the roles that HARMAN is playing today and will continue to in the future. The connection to the OEMs is indispensable. The latter, however, will no longer sell vehicles, but emotions. For an OEM, the key task will be to tune in to the personalised, on-demand, and situational delivery of content to ultimately create the optimum experience for users, today's drivers and tomorrow's mobility customers.

Barbara Flügge: Mr. Roth, can this be summed up strikingly with "from car manufacturer to emotion mediator"?

> Hans Roth: "Well, an OEM has to ask himself how he can control his offer in such a way that the emotions can be retrieved according to demand. When it comes to vehicle construction, extra space is needed in the interior, in the functional equipment and for connecting the vehicle to content providers in order to offer the right content via entertainment and infotainment functions – in order to transmit emotions and awaken new emotions in the vehicle."

Barbara Flügge: But this means that the OEM also has to take care of the digital control and equipment, and not just the physical, correct?

> Hans Roth: "Absolutely. It goes even further: "As an on-demand user, I will no longer be able to decide for myself what is offered in the car."

Barbara Flügge: Let me get this straight: I rely on those who make the race—depending on brand, location, or a combination of both?

> Hans Roth: "Exactly."

Barbara Flügge: Are we still interested in the brand?

> Hans Roth: "An OEM is thus also perceived for us on a different level. The brand value, i.e. the importance of the brand, does not take place via the number of horsepower and appearance of the car anymore, but via the emotionality it generates for us."

Our discovery journey with HARMAN and others alone show how digitisation effects given and well-moulded market and dominance structures. It is not without reason that experts describe the coming 5–10 years as the most important in the automobile segment since the invention of the car.

If we take a look at the explanations, for companies such as HARMAN and OEMs, the ability to bind to the end-user will be decisive. The end-user finds his way into the vocabulary as an on-demand user, mobility consumer, persona or simply customer. In connection with geography, other customer groups arise. Emotionality providers will further strengthen the relationship to existing customer groups, regardless of whether they are frequent or occasional drivers. Loyalty programs will add to the in-car experience room as a new service trade arena to their portfolio. At DoktorB, we have been working on new personas and customer groups, which we refer to as KPC: kids, pets, and cargo. KPC passengers and in-car consumers have not yet been able to use an unaccompanied vehicle, nor have they been holistically served by OEMs and service providers. We hereby exclude the traditional delivery modes and services offered by parcel service providers and self-conducted trains.

The scope of KPC may sound disrespectful, but autonomous driving will create new digital services to attract these new user groups, which are children, animals, and goods.

The children's pocket money, for example, will transform into a digital service funnel. Where they go, move, join, and spend will be covered in time slices and service proportions—preconfigured by their parents and those with custody rights. According to events, routes, and distances, the digital budget is monitored and accessed. In the context of pets, they can be taken to the doctor or be brought to a family member. The discussion about passenger and freight traffic to optimise traffic flow, utilisation of space, but also reduce noise and air pollution, is already taking shape in some pilot projects. Another group are the user group that we call OCN: occasional mobility consumers, coincidential co-travellers, and nomads will raise interest of future service providers for in-car offerings. Consider OCNs as, for example:

- Occasional: Occasional drivers,
- Coincidentals: Random passengers who are inspired by an experience,
- Nomads: People who live and work here today and there tomorrow.

Barbara Flügge: Mr. Roth, will we continue to find the brand-enthusiasts driving their beloved brand and only that?

> Hans Roth: "Of course! We find buyers or users of certain models or brands. We also retain the sports drivers who like to live out their speed preferences with additional functions. The latter two customer groups in particular, brand and sports drivers, continue to be car owners. While the previously developed customer groups, KPC and OCN, find themselves in the segment of MaaS users, i.e. the Mobility-as-a-Service users."

In the following chapter, we introduce more details about how HARMAN engages in innovation management.

4.5 Innovation Projects for the Car of the Future

For companies like HARMAN, market orientation and the consolidation of their own position in the market are indispensably linked to the demand for sustainable innovation. This is the only way for a company to develop and establish itself from a sound system provider to an infotainment provider, and to secure a place in the mobility value chain. Innovation projects have become part of the daily business.

Disruptive developments in the car and mobility environment demand special characteristics. This includes designing and doing. It is an "art to achieve an ideal composition of existing ideas and elements" ([4], p. 206). This is precisely because not every innovation makes it into the top 3 of the proposals that are presented to the board. Innovation management has a mandate to rethink in order to change perspectives and explore new market opportunities and new customer or user groups (see [4], p. 206).

What opportunities are there for companies like HARMAN to research and help shape the car of the future? Mr. Roth gives me an insight into two projects: the FANCI project and the cooperation with Rinspeed.

4.5.1 FANCI Project

FANCI was a project of the so-called Horizon2020 research programme of the European Commission [8]. The research assignment of the FANCI consortium was as follows:

> The FANCI Project will research, innovate, architect, design, develop and validate innovative multimodal face and body analytics and natural computer interfaces (NCI). It will develop a reference platform fitting a variety of applications and provide validation through user experience evaluation and benchmarking in the rapidly expanding automotive field. [8]

FANCI stands for "Face and Body Analysis Natural Computer Interaction", and the project had a duration of 28 months, from January 2015 to April 2017. Besides HARMAN, the companies CEVA, KeyLemon, tobii, NVISO, SoftKinetic and the University of Siena jointly researched the driver-vehicle interaction of the future. The aim of the two project phases was first to create an automotive demonstrator and then to create a universal reference platform that is not only applicable to the car. In the following, we concentrate on the automotive application.

Researching the foundation and mechanisms of a digital assistant go hand-in-hand with foresight research. The project investigated the influential factors from traveller and co-traveller on the behaviour of the vehicle. Influencing factors are the result of facial and body reactions, natural or physical interactions, such as braking or steering, as well as the expressed and/or perceived moods of the passenger or fellow passengers.

Experience spaces arise in two directions: (1) on the one hand, through early intervention and minimisation of hazardous situations and risk spaces, and (2) on the other hand, through steering the in-car emotionality and creating positive mood spaces. These are, for example, relevant for the selection of music, the information provided, and a setting of the digital assistant tailored to me as a driver or passenger. The constant training of the assistant is important here.

The project deals with these questions:

- Interactions between man and machine/vehicle
- Change in human productivity
- Car entertainment in general and personalised car entertainment
- Influence of concept studies on the development of human-machine interaction and its acceptance
- Influence on experiential and mobility behaviour through technologies such as artificial intelligence, new input mechanisms such as facial expressions, gestures, posture, and speech, as well as adaptive learning software and behavioural analyses

The project distinguishes twelve use cases in the focus areas of (a) investigations of the natural interaction (Natural Interaction), (b) driver evaluations (Driver Analysis), and (c) Safety and security (Safety) towards (d) personalised experiences (Personalised Experiences).

With respect to (a), Natural Interaction focuses on the control mechanisms in the autonomous vehicle (Mix Controls), an intelligent display and evaluation program me (Smart Dashboard), the digital assistant (Intelligent Personal Assistant), an intelligent navigation assistant (Smart Navigation Assistant), and intuitive multi-modal vehicle intervention (Intuitive Multimodal Adjustments).

Face authentication and usage-based insurance are on the agenda for (b) Driver Evaluations. Concerning (c), Safety-relevant applications include intelligent speed monitoring (Enhanced Speed Limits Detection) and a safety assistant (Smart Safety Assistant). For (d), Personalised Experiences, experience rooms in the car of the future, colleagues are focusing on operating new "driver groups", such as teenagers through Face Authentication for Teenagers Curfew, mood-controlled music experiences (Smart Music Players), and feeding user profiles into the car (Identification and Loading Profiles).

On-Off Experiences

In the future, we will design experiences in a vehicle according to individual taste and create our own experience based on our own taste. Thanks to a sophisticated digitalisation and preference control, experiences will be tracked, composed, and offered within a few seconds. Herein, they stimulate our decisions towards an experience (or we could reject another one). Cognitive imagery emerge from a simulation of multiple images we mentally and physically collect—with these being our mood, the environment, what we see, hear, smell, and/or our appetite for adventure and experiences. These imageries serve as a switch and determine how we are. A change of scenery, a new interior, or an upcoming visit will become mini-experiences in an unprecedented environment.

The FANCI project team has set itself the goal of translating the findings of the use cases into a conceptual description of the so-called FANCI platform. The reference architecture serves, on the one hand, as a development platform and, on the other hand, for load distribution. For the former, a so-called FANCI API is made available, which, as a multi-modal interface, gives developers access to the testing and transmission of new information channels such as facial expressions, speech and gestures. These information channels can also be found under the term "pervasive natural computer interaction devices". In addition, with regard to sharing the digital workload, the project consortium is concerned with being able to distribute the workload and activity requests to the mobility user according to the situation and preference. In ◉ Fig. 4.1, the solution components of the FANCI platform are transferred into our Smart Mobility reference architecture.
Barbara Flügge: Why does it need a platform approach, Mr. Roth?

> Hans Roth: "A contextual reference or recommendation to the consumer cannot always be available in the cloud. The aim is rather to provide reduced functions as well as basic functions of the personal assistant even with a lower bandwidth. This means that the colleagues are researching the offer components of the digital assistant once off-board, i.e. offline, and once on-board, i.e. online. The role play between man and machine is playing a role in this today and in the future. The latter is required, for example, for the question of to what extent a vehicle may, must and can intervene and to what extent a vehicle may override human preference – the above example of weighing the importance of a telephone call and the resulting lack of attention to 100% concentration in a hazardous situation."

Speech recognition is used to filter out and control recommendations and services. Those who change from car to motorcycle, or from car A to car B, experience a transformation of the consumer experience. We call it the transferability of the consumable desire from one vehicle to another. It is one of the core functions of the so-called individual mobility account that manages our digital identity per individual customer. We define it as follows:

> Any individual mobility account is pulling in operational functions such as identification and authentication, payment, reporting and the configuration of personal trust gates. [...] Operational tasks are conducted purely digitally and via the cloud. Always accessible, the mobility account drives the configuration of my mobility experience and needs and can be applied to distinct means of transport – shared and owned, be it in the autonomous vehicle, the vehicle from the company fleet, or the vehicle just booked from the shared mobility network of my municipality. ([4], p. 107)

In addition to the introduction of the FANCI project, we have been looking into the topic of road safety in the context of autonomous driving [9]. Dr Tscheligi and DoktorB have investigated the question of value creation through networking and self-learning systems and the importance of human-machine interaction (HMI) in the car of the future at the Business Lounge event of the Automotive Cluster AC Styria in Austria.

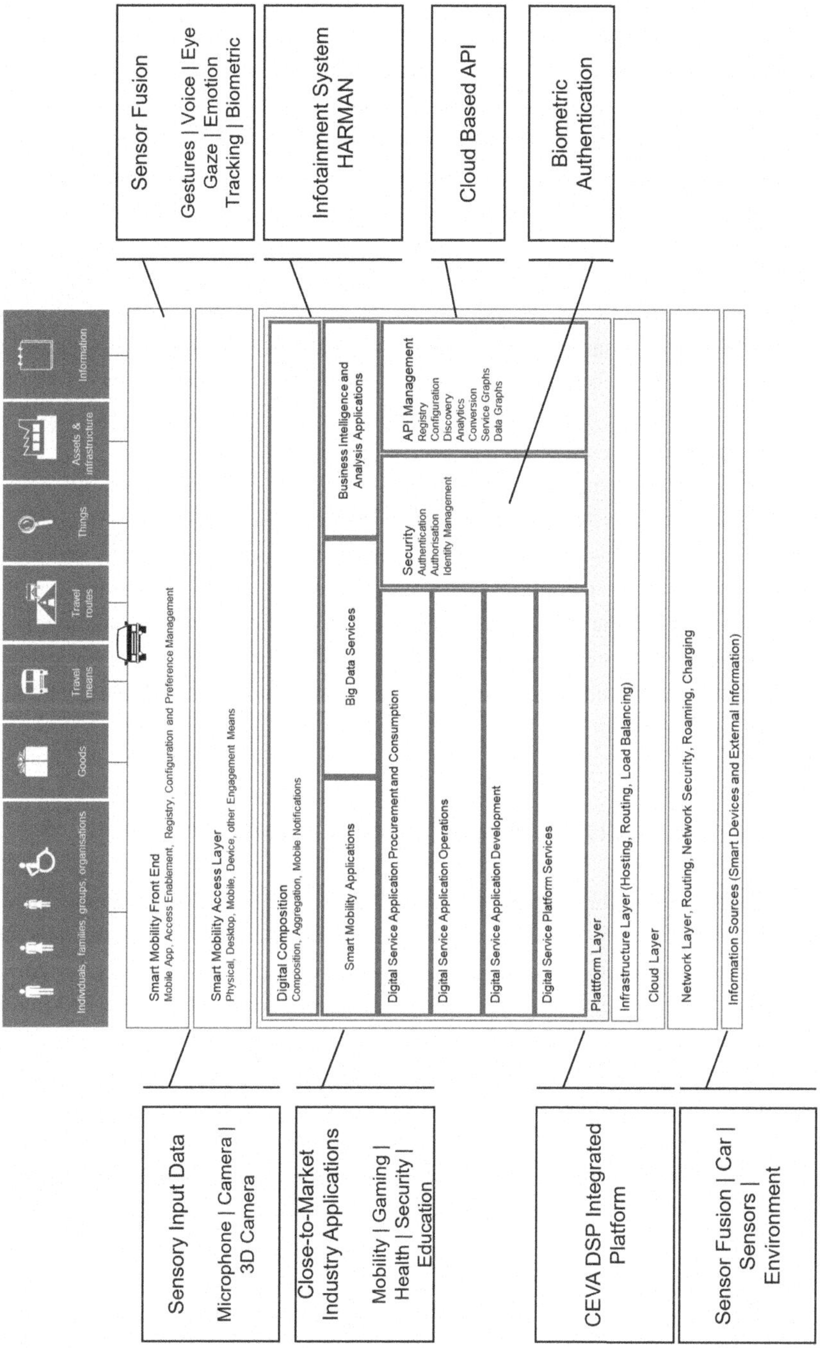

Fig. 4.1 Embedding the FANCI platform in the Smart Mobility Reference Architecture

Dr Tscheligi talks about the history of actions and reactions that are built within the vehicle. This history, so Dr Tscheligi states, influences the in-car configuration, and steers the self-learning system. We herein are concerned about a conflict raised by a driver's preferences and actions emerging from a vehicle's preferences. What situations steer a driver's decision to intervene early and what situations make the driver wait longer? Simulating and configuring the reasoning of an autonomous vehicle will turn into a multi-year, if not multi-decade, effort. We would like to add here that we need simulations to find out when rationale and when an individual preference have an impact on driving behaviour and mobility offers.

As with many companies, the FANCI project is a great contributor to co-innovation in a collaboratively and thematically relevant environment. We ourselves work(ed) alongside projects with OEMs and technology providers such as IBM Watson and other AI providers.

HARMAN, for example, is gaining further insights into the demand of the future. It is the persona-specific studies that serve the co-creation of new mobility services—compared to traditional market research.

In our services concept we completed for NEMO 2.0 (NEw MObility 2.0), we demonstrated how to design the persona and journeys for smart mobility and experience-enhanced service designs for rural areas in Germany for providers, consumers, farmers, businesses, and consumers of any type—truly holistically. Use cases for managing the mobility journeys of teenagers and scholar groups were included, too. Those look into managing the rules for in-car service consumption. These rules are set by parents and cover, for example, the consumable budget and car sharing rights to book certain routes and visit events like concerts and other entertainment.

4.5.2 Innovation Mediators: From Concept Cars and Rinspeed

We cannot imagine the world of the automotive industry, CES Las Vegas or the Geneva Automobile Salon without design studies and tangible future car prototypes that allows to touch and see model novelties. Irrespective of geography and brand diversity, we always like to be inspired by the ideas of creative minds. DoktorB herself was invited to join the Daimler design project called "the design:project" in Budapest—as part of a pan-European design and user experience initiative. The design:project was concerned with the assessment and evaluation of the market-ready GLK model. The assessment was concerning the experience as driver, co-driver, passenger, bypassers, artist, design afficionado, and experience seeker. Not only have we tested the car but we worked on its technological, functional, and design-driven impact to the urban ecosystem. The project gave Daimler the opportunity to refine their service offerings along the design-to-deliver value chain of a new model.

Other than the design:project, we have also worked with concept cars and herein we evaluated HARMAN's and SAP's strategic role and tactics in the context of Rinspeed.

With respect to concept cars, these play a role long before a car is ready for series production—including the necessary ecosystem that consists of the OEM, suppliers, assembly lines, and many others. And honestly, the concept car might never be intended to be a serial car.

Interviews and surveys do not play as large a role in HARMAN's innovative design as the design thinking workshops and the selective involvement of end customers play. It is the concept car that comes to the fore as the pivotal point for thought leaders and innovators at HARMAN.

In order to present itself as a technology company and continue to exist as a thought leader, companies like HARMAN are interested in a central medium that can be used for futurity and innovation research. A concept car translates a company's claim into a haptic, tangible structure. Thus, ideas and development progress can be presented in a concept car and are virtually visible on the living object to the company's own divisions. They are also more comprehensible than on the computer screen or in an isolated simulated structure. Discussions about size, accessibility of a control panel, optics, and equipment can be conducted much more purposefully in a concept car than on a digital drawing board.

A company that participates in a concept car presents itself not only as innovative. The company becomes a pioneer and thus offers a promise for the future to its business ecosystem. A promise that speaks about the mission ahead of time and how a company will be shaping the future. If a stakeholder is progressively acting, it is easier to ideate the future. Customer acquisition and business development are just as much a measure of the success of innovation as visibility in the company's own business network.

> "HARMAN", says Hans Roth, "presents itself as a forward-thinking supplier to its customers such as OEMs through the concept car. If the concept car goes on a journey, the focus is on those business partners who are located further away in terms of geography or value added."

A concept car is also a decal of how the future will present itself in a car and how the future will offer itself to us, the drivers, and passengers. The concept car thus becomes a mediator between those who wish to offer something and those who will in future demand the services offered or envisaged. The concept car translates, so to speak, the future for us and lets us participate in how innovation can be lived. This of course presupposes that we get a chance to see and experience the concept car at all.

This is where the 10-year collaboration between HARMAN and Rinspeed comes in [10, 11]. The partnership with Rinspeed offers HARMAN decisive advantages in order to make a contribution in the field of autonomous driving and to present itself as a technology company. The signal "with HARMAN into the future" does not miss its effect.

> Mr. Roth describes it this way: "An OEM sees the supplier as a forward-thinking business partner. In addition to the reflections of customers, who can see the car in the first place during its journey around the world, engineers and technicians of the OEM can also deal with HARMAN's innovations to a much greater extent.
>
> We don't have to wait for specifications from the OEM, but can take care of the technical embedding of our systems in the vehicle much earlier," Mr. Roth continues.

If it goes back to the design process of an innovation, a technical detail or an extended assistance functionality in the infotainment system, other aspects can suddenly be taken into account in the context of the car: Interior, sound, operability. It is also noticeable that the demos, as we know them from product presentations, are usually limited to one specialist area: the concept car, on the other hand, offers itself as a presentation platform for a wide variety of areas!

Rinspeed is all about partnership and cooperation between partners and technology companies [11]. Each company can bring its latest technology to the vehicle and market it. From plastics manufacturers, fabric manufacturers, experts in lighting technology, to solution providers such as SAP®, the vehicle is packed full of innovations from different business areas.

The Rinspeed company is responsible for coordinating the vehicle's appearance and focus. Possible conflict potentials are solved together. If there are potential competitors for a concept car or model at the same time, these are either integrated into the vehicle in the sense of a co-existence, or one decides to be there or not. There is enough "space" in the car to show variety and position yourself.

A further aspect is the following: there are integration points, and even if "only" playfully staged for a car, connections, and dependencies between entertainment and interior equipment, the operability of one system and the sequence of routes in the other system can be identified. The first stage of the confrontation is already done. A further interesting study is the follow-up projects between two or three companies that have come together on the basis of their cooperation with Rinspeed.

The coordination process, which is geared towards the target image of the car, ensures success among the partners, both in the concept phase and in the market presentation. Measured against the costs normally associated with concept cars, it is not excessive to invest a single-digit million sum for participation. In regards to the rights regime, the rights to the marketing have been acquired by the partners such as HARMAN and the publication rights are regulated by Rinspeed [12].

In the coming years, we will see to what extent the focus will be on the multimodality of in-vehicle services and user profiles. The different modules will then turn the car into a meeting room or a mobile home. If we look at our mobility and service needs on a working day or weekend, we can quickly decide what is important to us and when. To see this depicted in a vehicle complex will convey the role of the car as the experience space of the future even more strongly than before!

The Future Matures in the Concept Car
Sustainable and absurd ideas consciously find a home in concept cars. Unbelieving amazement, discussions about feasibility and taste, all this is needed in order to give innovative strength room for manoeuvre. Concept cars as the result of a collaborative innovation process are also mediators. They mediate between the worlds of illusion and reality. Ideally, innovation selection processes will take place in one vehicle in the future—both with those who want to offer something and those who can't even imagine the offered or planned services!

4.6 DoktorB's Checklist

Important Basic Considerations

- What experiences would you like to share with your car?
- Which of your services can be offered well in a car—be it for information, for ordering, for evaluation?
- Where do you see organisational hurdles to consuming services in a vehicle instead of in a department store or boutique, in a gym or at the airport?
- How do your sales and marketing expenses change with the addition of cars as a new, omnipresent marketing channel?
- How well do you know the wishes of your customers?
- Who will be your competitor in the future if you use the vehicle as a sales and marketing channel?
- How can even complex services be advertised in a vehicle?
- During a holiday trip you often use different means of transport, including a rental car—should the vehicle recognise you? And if so, what should the vehicle know about you?
- In your environment you will find business partners who have grown with your company? How do you prepare your business environment for new digital offerings? How much does the car become a bottleneck?

References

1. Emily Cretella (2013) Anthrophologie's fabulous approach to content marketing, published on RAGAN's PR DAILY on 11.10.2013. https://www.prdaily.com/Main/Articles/Anthropologies_fabulous_approach_to_content_market_15343.aspx. Accessed 17 Feb 2018
2. HARMAN International (2018) Company website HARMAN International. http://www.HARMAN.com/. Accessed 17 Feb 2018

3. Audi (2017). http://www.audi.com/ci/de/guides/corporate-sound0/corporate-sound.html. Accessed 17 Feb 2018
4. Flügge B (ed) (2016) Smart mobility – trends, concepts, best practices for intelligent mobility. Springer Vieweg, Wiesbaden
5. Website Waze (2018) Company website Waze – get the best route every day with real-time help from other drivers. https://www.waze.com/de. Accessed 17 Feb 2018
6. Website Rinspeed (2017) Company website. https://www.rinspeed.eu/de/Oasis_21_concept-car.html. Accessed 17 Feb 2018
7. Echo-Tests.de (2018) Alexa – who or what is Alexa and what can she do? https://www.echo-tests.de/alexa-wer-oder-was-ist-alexa-und-was-kann-sie/. Accessed 17 Feb 2018
8. FANCI website (2017) Project website. http://www.fanci-project.eu/. Accessed 17 Feb 2018
9. AC Styria Website (2015) Event announcement Business Lounge Car Interaction Safety & HMI. http://www.acstyria.com/_files/Events/Events_2015/BL_HMI/151009_ACstyria_HMI.pdf. Accessed 17 Feb 2018
10. Website HARMAN (2017) HARMAN unveils the ultimate in-car experience for intelligent, intuitive autonomous driving at CES 2017. https://news.HARMAN.com/releases/HARMAN-unveils-the-ultimate-in-car-experience-for-intelligent-intuitive-autonomous-driving-at-ces-2017. Accessed 17 Feb 2018
11. Website HARMAN (2018) Rinspeed and HARMAN adapt to driver needs in a SNAP. https://news.HARMAN.com/blog/rinspeed-and-HARMAN-adapt-to-driver-needs-in-a-snap. Accessed 17 Feb 2018
12. Website Rinspeed (2018) Company website. https://www.rinspeed.eu/de/. Accessed 17 Feb 2018

The Digital Living Space

5

DoktorB Barbara Flügge and Hans Roth

Abstract

The marketing billboards tell us great stories about the exponential shopping experience in a whole new world: the yoga class in the smallest spaces while we are on the move, the ultimate quiet zone and mobile power napping. Innovators and visionaries like to expose us to a variety of ideas. But what about their implementation? Isn't it just while we are on the way to making a decision that we discover what, when, and how will work best? We often experience the flaws of online connection, are restricted by our constantly updating calendar, or we even encounter that our favourable piece of clothing, which has always been too expensive to become affordable, is suddenly offered at a cheaper price at the boutique around the corner. Imagine, that in the near future the car will give us an ever greater parameter of activity, both physical and digital.

D. B. Flügge (✉)
digital value creators (DVC), Bottighofen, Switzerland
e-mail: barbara.fluegge@dvcconsult.com

H. Roth
Harman International Automotive Division, Karlsbad, Germany
e-mail: hans.roth@harman.com

B. Flügge (ed.), *Smart Mobility in Practice: Good Car, Bad Car, No Car – Is this the End of Nurturing Our Mobility DNA?*,
https://doi.org/10.1007/978-3-658-43171-6_5

5.1 Anytime, Anywhere in the World

Remember the scenario and examples of HARMAN we introduced in the previous Sect. 4.3. With regard to social fashion, we drew the picture of the webshop in the car, which, for example, offers us price comparisons on our favourite coat with pictures and background music, and asks our best friends in New York and Zurich for their recommendations. Since we start the query in the evening, this fits well with the local time zone in New York. If we had wanted to inform ourselves about the feedback of our friends in the morning, the girlfriend in Zurich and the girlfriend in Vienna would have been invited to chat. Is this utopia? Certainly, this scenario can only be imagined and implemented by the means of a digital concatenation of individual solution modules. But this is not utopian. Tesla has long been demonstrating to the classic automotive industry what is digitally retrievable and consumable, and how a new functionality can be installed with a single click in the car while moving.

What is the benefit to Tesla drivers and passengers? Although consumer and mobility services are not always cheaply prepared, they are skilfully and attractively offered to drivers and passengers. Regardless of whether we can afford a Tesla ourselves, ride along, use a local taxi service like in St. Gallen, which has a Tesla in use, or book a Tesla in the Mobility-as-a-Service package offered in Oslo, the Tesla users know more! The advent of digitalisation gives customers of disruptors like Tesla a head start. In this case, users and customers of the disruptor participate in the digitalisation journey and are being educated all in one. Here are some learning examples: How does online booking work? How do I experience innovation? What ideas can I carry into the workplace or a project as I experience disruption myself and become a part of an innovation system?

Speaking of innovations and experiencing distinguishing features plays a key role in the consumer market. This first consumer advantage becomes a great talking point, making us proud and encouraging us to share pride in the product. Speaking about innovations is one side of the coin, but who consults and implements? Shouldn't consulting firms that are committed to digitalisation earn an armada of digital experience worlds in order to have a say?

How do we manage to involve those citizens who can neither afford a mobile experience studio nor a Tesla driving service and invite them into co-creation? Mobility should be affordable for everyone. In a workshop conducted by DoktorB, we had the pleasure of co-creating with a governmental representative from the city of Atlanta, Georgia. They inspired us with their story of the City Design Studio. The City Design Study looks like a small shop with shop windows and opening hours, but everyone can go there and hand in their ideas, suggestions for improvements, what they like, and what they experience in the city.

Would you, our dear readers, join us and open a Mobility Design in your town, village, or even digitally? And would you be willing to offer affordable experience moments that we establish jointly in your vehicle? There should be a place for it with the increasing vacancies in the retail business. And as far as the Teslas and other disruptive vehicles are concerned, an experience area will become accessible to all and become accessible without being concerned about affordability or other social thresholds. The issue of finances should be reduced and knowledge should therefore be skilfully conveyed. We highly recommend that cities and municipalities should consider how to engage their citizens in digitisation and innovation for city centres that are fading away.

Let's get into our car. To us, the car becomes a digital living space. Even today, vehicles without IT and telecommunications services are no longer conceivable and cars depend, to a large extent, on their functionality and well-established upgrades. There are unresolved questions about the business models of the future and how today's and tomorrow's vendors are dealing with network effects and growth forecasts. We will be made even more aware of how much diversity is worth to us.

5.2 Business Models of the Future

Who pays for mobility and who for the ultimate experience? Depending on the purpose, different price models become available. However, if we are the ones who leave behind data traces and equip a service provider with our data, it would be obvious in terms of shareholder value that we would also be remunerated. After all, many providers benefit or solely exist because of our data submissions and contributions—whether voluntary or unconsciously.

Let's have a look at the different forms of mobility-enhanced experience. We distinguish experiences as follows: self-determined, co-riding, and co-consuming (Fig. 5.1).

- Self-determined means that we decide for ourselves what share and content of our mobility and consumer behaviour we will communicate, sell, and expand with a third party experience. The central question here is: "do I really have to pay money or is it enough if I share certain experiences?" Self-determined experiences are driven by our choice and our consent to share behaviour and insights before, during, and after a trip and to what extent we share them.
- In the case of co-riding, the passengers consume, and benefit from, the experience determined by the driver of other passengers as to what is being offered in the vehicle. Think about news, podcast episodes, recommended retail items, or a travel guide. The passengers are clearly not paying for the experience.
- In the case of co-consuming, we envision co-consuming in groups of like-minded people. These are, for example, technical experts who receive training while traveling to a conference by vehicle or travel module. This might be on-site training on product innovation or a briefing on an upcoming business appointment, or even the preparatory session that informs about the prospective upcoming meeting.

Any of the three experience forms can be combined.

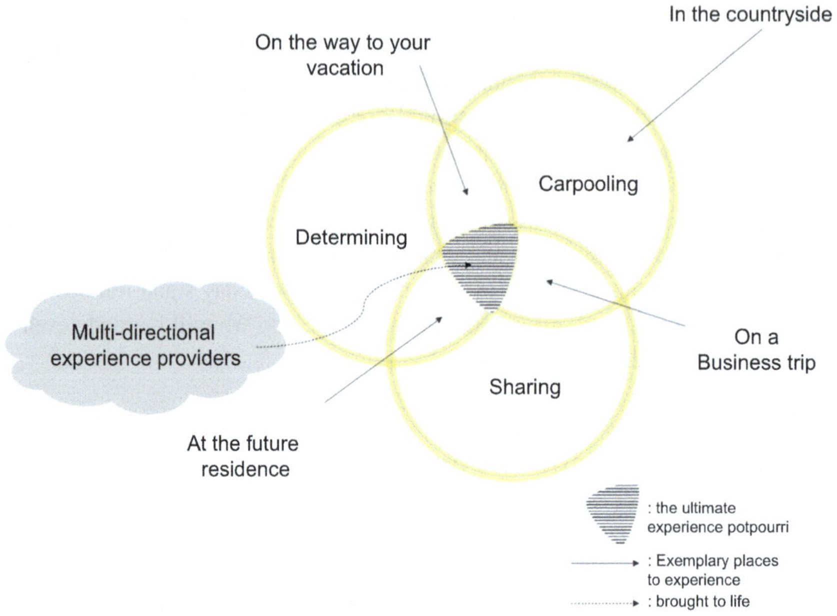

Fig. 5.1 The business model of the future: The consumers' favourite points of view, Consumer view of the experience, spaces of the future

Any of the three points of view are adaptable to the local context and the position of the vehicle:

- Location being a city
- Location being the countryside
- Location being a company's headquarter and its vicinity
- Being on the road to a holiday destination
- Being on a business trip
- Being elsewhere, e.g. moving to a hotspot

If you are part of a project team that is currently working on ideating the first suite of experience services, take a look at your region or the targeted geography. The geographical alignment helps you to design the personas, the desired experiences and presentation layers, and the characteristics of the space itself. All of the characteristics of the location are put into relation with each other. We run these kind of workshops in a fairly overseeable number of hours based on our experience and prior projects.

One question that arises here is who will pay service engineers like digital value creators (DVC) Switzerland and experience providers like HARMAN? Those who manufacture vehicles, models, and devices will certainly continue to seek direct contact with the end-customer. Efforts are also being made to attract new groups such as KPC and others.

Premium services are the main focus offered to your install base. Further services can be displayed via service portals [1]. If OEMs become service brokers, with two options, being it their own brand image and secondly a whitelabel solution. HARMAN and its partners could opt for the second variant, transform and expand, becoming a service aggregator. Smaller OEMs, or OEMs on small scale, apply selected brand-related services. In particular, when it comes to differentiation in the market, direct marketing is an obvious option for certain customer segments. Nearly 100% of service requests and service distribution, the so-called brokerage, take place via the OEM. How this is illustrated can be seen in Fig. 5.2.

With the spread of MaaS and the concept of the holistic mobility management of a city, the smart city will in future act as a distributor to end customers. If the city delegates the offers to carsharing or MaaS providers, for example, it is they who compete for the end-customers' favour. Whether one variant or another, the service landscape changes in all cases. With a multiplier like a city, even service providers like HARMAN can move up to the end customers and thus bypass the OEM, for example. In terms of structure and process organisation, it takes a lot to serve the new clientele. Take a look at our previous projects with regard to customer service, service hotlines, and setting up a service-focused sales organisation.

If a MaaS provider engages in a city and collaborates with that city as a multiplier, it controls the city's fleet offer (see Fig. 5.2), which consists of a mix of public, private, and shared modes of mobility. This could be, for example, a mix of public transport, taxi, car sharing, bike sharing, rental cars and modern forms of locomotion, such as autonomous cells. The dependency or variety of services then takes place in a regulated environment by the MaaS provider. In Fig. 5.2, we have presented a variant in which autonomous vehicles

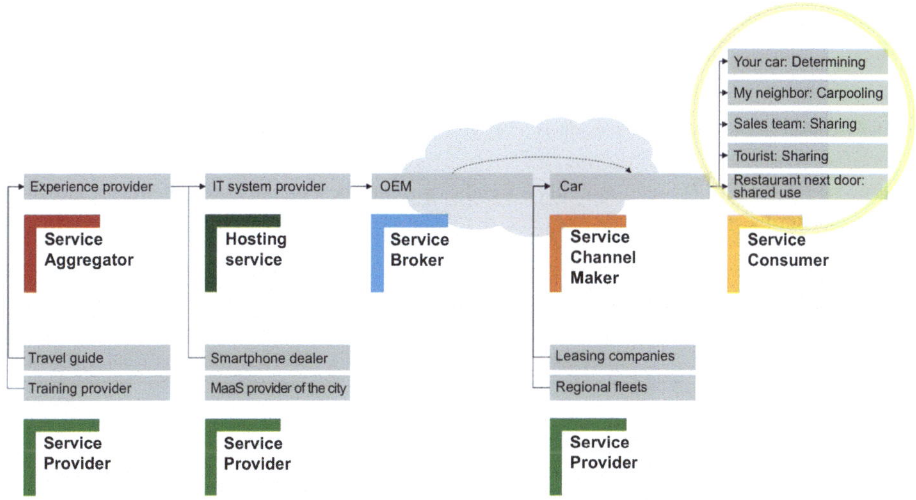

Fig. 5.2 The IoS role model for adventure spaces

assume the role of channel makers—regardless of their operators, vehicles are perceived by consumers as direct business partners. The vehicle and consumer then decide on the conditions of the experience purchase and benefit. The initiation of business by means of a mobility medium will radically change the market. This happens to the extent that service providers can only maintain and, if at all, expand their position in the market through their deliberately multi-functional, yet experience-focused and personalised offering.

5.3 DoktorB's Checklist

Important Basic Considerations

- Which services could you offer at any time to the citizens and companies in your city or district?
- You are a provider of transport services: How much do you know about the digital needs of your customers? With more information provided by users, could you offer other, better, or more targeted services?
- How do you assign the experience forms "co-determining", "using" and "driving along" to your customers, your business partners, and your desired customers?
- What would you have to do to bring about a transformation to digital offers in your company?
- How high is the share of digital services today, measured on the one hand by the range of services you offer in the market and on the other hand by your product range?
- How much digitalisation can your company afford today and how much will it cost in the future?
- On what calculation basis do you want to use, refine, or distribute data in the future?

Reference

1. Flügge B (Hrsg) (2017) Smart mobility – trends, concepts, best practices for intelligent mobility. Springer Vieweg, Wiesbaden

Surviving with the Car: The aCar

6

DoktorB Barbara Flügge, Martin Šoltés, and Sascha Koberstaedt

Abstract

On one of our trend scouting tours we discovered the small booth of Sascha and Martin and asked for a pitch. During the first 15 min, I realised that this was not only a cool research project, but a vision with a clear mission and solid foundation that would be worthwhile exploring further and ideating a future scenario. The project of the Technical University of Munich (TUM) and its project and support partners has proven how sustainable mobility can be shaped and implemented in developing countries. Securing the survival of families in rural or remote areas triggered the go-ahead for an unprecedented project. Criteria such as security, supply and income security, as well as the establishment of new value chains, control the innovation process. The innovation, the aCar, presents itself to the manufacturing and automotive industries as a multi-functional, modular and mobile service provider.

When we talk about autonomous driving, we mean an object, in this case a vehicle or a module, which without a driver can independently master all tasks in all situations. But autonomy also means in the broader sense to be free and self-determined, and to attain a degree of self-administration. In particular, the latter definition of autonomy is experienced by the farmers of Ghana through the aCar. The aCar [1] is a very special

D. B. Flügge (✉)
digital value creators (DVC), Bottighofen, Switzerland
e-mail: barbara.fluegge@dvcconsult.com

M. Šoltés · S. Koberstaedt
EVUM Motors GmbH, Garching, Germany
e-mail: soltes@evum-motors.com; koberstaedt@evum-motors.com

B. Flügge (ed.), *Smart Mobility in Practice: Good Car, Bad Car, No Car – Is this the End of Nurturing Our Mobility DNA?*,
https://doi.org/10.1007/978-3-658-43171-6_6

vehicle, which was developed through the cooperation of research and foundation institutions in the USA, Nigeria and Munich under the direction of the chair of Prof. Lienkamp [2]. aCar is also the history of EVUM Motors [3].

6.1 Everyday Life in Ghana

6.1.1 Autonomy Without a Car?

Let's go on a journey to Ghana and imagine a dusty road that has been entered onto the road map purely because it provides access to the main road, and not because of its condition. In 2017, Ghana had 28.8 million inhabitants in an area of 238,537 km^2 [4]. For comparison, Ghana is only as big as Great Britain. It's infrastructure is characterised by roads and, thus, also indicates the people's preferred mobility infrastructure. The share of rail and air connections is comparatively low. The region that we targeted in our trend-scouting visit is named Ashanti. Ashanti is home to five million people. Kumasi, the second largest city in Ghana, has about two million inhabitants and is the main city in Ashanti. The rural regions are characterised by a number of villages, where 10–20 families live. The families often have four to five children and use a small piece of land for cultivation. Cooperation between several families is not uncommon. In addition to small farmers, there are also large farmers who, for example, have large orange plantations.

Let us imagine a country in which survival in the countryside is oriented towards today. Focus is not on the anxious questions of how competitive the produced gifts will be, whether they will make it to their destination in time, or whether it is a question of commissioning a drone to deliver them or preferring the traditional route via the parcel service. Martin Šoltés and Sascha Koberstaedt, the project managers of aCar and the founding fathers of EVUM Motors, have been asked the central research question of how to ensure the ability to survive in rural areas.

Everyday life in Ghana is marked by mobile inequality. In urban areas, used cars that have reached the end of their useful life and would earn just 500 euros in the European market are traded and sold for no less than ten times as much, i.e. 5000 euros. Whoever can afford it buys one. Ownership in rural areas is scarce. Whoever owns land also owns a house. Buying and owning a vehicle yourself can be found among the large landowners. Vehicle buyers are rarely users of the vehicle, instead their profitability is triggered by renting out the purchased vehicles. If a family can afford a vehicle, it is made available through carsharing within the community or renting it out to family and friends.

In rural areas, the survival of the community depends on functioning connections with the next largest cities. To date, privately initiated carsharing and bus-sharing offers have been available for this purpose. These are expensive. In addition, the structural connection of the main road to the villages is faulty. The inadequate or partly missing infrastructural connection influences the possibilities of income generation for farmers. If workers have not moved from the village to the city, they have no choice but to cultivate their land. Those who own the land cultivate it.

A core product of agriculture is the cassava plant. The cassava plant, often referred to as manioc or simply cassava, is also known as yuca. According to food scientists and nutrition experts, it is one of the "most important food plants in the world [...] and represents the fourth most important source of calories for the tropical population after rice, maize and sugarcane and provides food for more than 500 million people [...]" [5]. According to the experts, the plant is "despite its importance [...] mostly cultivated by small farmers for their own consumption or the local market" (ibid.). Since the plant has a high prussic acid content and spoils quickly after harvesting, it must be processed as quickly as possible or sold on the local market. Further processing and fermentation produce, for example, pasta, so-called Agbelima, or Attiéké, a product similar to couscous. You can also process flour, soups, or gari porridge.

The farmers receive too little money for the direct sale of manioc on the spot. They aim to trade the fruit in the next biggest town. For the distribution of their products, farmers depend on predetermined mobility offers such as carsharing or paying for a share of a truck loading area. Those who operate trucks coordinate with each other. They push up the price of transport or loading areas. Even if the farmer had the financial means to hire a lorry, he would have little chance of getting a lorry because of the prevailing monopoly of truck owners.

For the transport to the city, a farmer has to mark his plants. Several farmers share the loading area. If the farmer does not mark his product, the goods cannot be demonstrated to be his. The goods are often damaged or disappear during loading and transport. In an ideal and smarter world, the truck should be multi-functional so that goods can be clearly assigned, safely stored and easily accessible. If the transport to the city takes place, the farmers follow the truck in a public bus. They have to pay for the bus ride, too. If the farmers shared a taxi, for example, they would be left behind. The public bus has a right of way in all matters.

If delays occur during transport, or loading and/or unloading takes too long, the farmer will not be able to sell his goods on the city market in time. Every delay costs money and effort. The product spoils quickly in its raw state. This is especially true of the cassava plant. In order to enable families and village communities to continue to stay in rural areas and be able to build/establish a social life, mobility is key—personally and for the product trade to achieve income for the families. There is a need for quicker connections and the immediate delivery of products to the city. This situation applies to 90% of the target group in Ghana as well as in other African countries.

Prof. Lienkamp's research team, led by Martin Šoltés and Sascha Koberstaedt, quickly realised that their altruistic goal of ensuring the survival of rural communities could only be achieved through a creative approach. And that it needs a vehicle! A vehicle that meets a wide range of requirements and makes use of technological, environmentally compatible and affordable innovations.

> **Focus on Service** The car could be your survival space, your escape room—as a multifunctional device it could serve for the transport, cooling, and storage of raw materials and semi-finished products, for power supply, safe locomotion and as a service provider. The car therefore joins the value chain for the farmers and their transformation in becoming service providers and brokers.

6.2 The Vision

6.2.1 Owning a Car: Just Any Car?

Many people have the dream of owning a car. From the point of view of the urban (rather than the rural) population of Ghana, this is feasible. There, as in large parts of West Africa, old, disused vehicles from Europe can be found. As mentioned above, old, demolished, and other sorts of secondary market cars find their owners in Ghana for ten times the European market value. The purchase of a car is driven by purpose and serves to generate income. Only the upper class, about 2% of the buyer's market, sees a vehicle as a status symbol and invests accordingly. Ghana's mobility profile is shaped by vehicle mobility. There are few flight connections. Passenger and freight transport generally takes place via vehicles.

The vehicles are used effectively and mostly during the day. Vehicle owners and users expect a feasible range of 80 kilometres per day. Cars, if available, would not be left unused in the garage or at work as they are in the European or US market.

In terms of brand diversity, a few brands have established themselves. Where you find an automobile brand, it could dominate the market. Brand dominance drives the availability of spare parts and where and how spare parts can be traded and purchased. Furthermore, it is key to local buyers that their local workshops gain experience in dealing with repairs. In Ghana, there will be no brand change in this respect, as it is same brands of old vehicles that are mainly imported and used from Europe. Through this system, the Opel brand has become widely used in Ghana.

I wonder if sustainability is an issue. When importing vehicles, the import tax is changed accordingly on the basis of the vehicle age. In this way, up to 100% or more of the purchase value of the vehicle plus additional tax can be collected. According to Sascha Koberstaedt, there is motivation for environmental protection. This is because, for example, during an oil change at a workshop, the oil is let directly into the ground, meaning an owner can literally see this environmental damage. Additionally, there are disputes about resources: who owns the resources and who wastes them? Solar power, on the other hand, is free of charge and has been used for some time via solar panels. In contrast, gasoline is a big investment. In addition to pricing, points of acceptance play a major role for the rural population in particular. Where no infrastructure is available, a distribution network does not exist nor is worthwhile either. The installation of such an infrastructure set-up is costly and herein would call for a sponsor to make it happen. Other environmental protection issues, such as particulate pollution, are not currently seen as an issue. This will certainly change with increasing prosperity.

6.2.2 What Do We Need for an Innovative Mobility Concept?

The research team had set itself an altruistic goal: to ensure the survival and growth of the rural population. The implementation required a vehicle structure that takes into account the everyday life of the rural population in general and the farmers in particular.

The vehicle should meet the following different requirements:

- Sustainable and cost-effective drive
- Selection of a drive variant that can be implemented on land
- Low-maintenance vehicle

The choice was made pretty quickly, and the team opted for an electric vehicle. This was not so much because of environmental aspects, but practical factors. The environmental considerations and decision-making to opt for electric mobility is currently more of a concern in Europe or North America. Nevertheless, the vehicle should also aim to avoid any potential environmental risks, such as air pollution. The costs were given priority: the vehicle had to have low operating costs, low-maintenance, and be reloadable everywhere. Electricity would be transmitted from solar panels, partly because it would not have been possible to transport gasoline to the points of destination, such as rural farms. In addition, Western vehicles have complex electronic controls. What was needed was an engine drive that would be cost-effective and that could be processed, operated, trained and used with few resources.

The aim of Šoltés and Koberstaedt was therefore to exclude combustion engine-based vehicles from the outset and to switch directly to an electrically powered vehicle. From a technical point of view, according to Sascha Koberstaedt, this is the best option; an electric motor has a very high starting torque directly from standstill and is therefore very suitable for bad roads with a required climbing ability. Figure 6.1 shows the road conditions on site.

In addition to the questions on the operational capability of the vehicle, further spatial and safety-related aspects were added due to the local context. The vehicle had to offer theft protection, not only for the vehicle itself, but also for the goods transported in it. Safe loading and unloading was required in order to a make work easier for the users and b. protect the goods from damage. Freight insurance is not an option today: it is considered a levy on the state and is not in demand. There are also no car or household insurances, nor liability coverages to be insured.

Any form of online instruction and learning material, such as YouTube videos, should become the primarily learning hotspot and educate about the support and maintenance of the vehicle. Wi-Fi and mobility network access is not an issue. According to Šoltés and Koberstaedt, the mobile network in Ghana is much better developed than in Western countries such as Germany!

Fig. 6.1 On the country road. (Courtesy of the Technical University of Munich/Mobility Systems Engineering)

6.3 From Vision to Execution: The aCar

The result of a 36-month research project is impressive: an electrical all-rounder, according to Prof. Lienkamp, that the team called aCar [6]. The research project started with the aim of investigating the mobility needs in rural areas of sub-Saharan Africa in the years 2020 to 2030, according to the tenor of the project team. He continues: "The overriding goal is to develop a coherent vehicle concept and thus to address the problems of the rural population with regard to the mobility of people and goods. The vehicle will provide better access to healthcare, education, and information in remote areas through digital technology that can be integrated in a modular way". Originally called "solar car", the local farmers opted for aCar, a name which consciously leans of the ideas of "any car"—a car for everyone. We will now depict the individual project phases. At the end, we add DoktorB's own projection for turning the innovation project into a serial production.

6.3.1 Project Phase I: Market Research

Starting with the first idea in 2013, a small project team was formed. Sascha Koberstaedt and Martin Šoltés are still in charge of the aCar project today. For 3 years, 100 students and 20 doctoral students from different disciplines and research institutions worked on the implementation of the vision.

In addition to the initiators of the TUM and African universities, the Rosenheim University of Applied Sciences, with its focus on raw materials management, and the Institute for African Studies at the University of Bayreuth were on board [6]. The industrial partners were [7]: the African Health & Agricultural Foundation, the automotive supplier DRÄXLMAIER Group, the manufacturing specialist Hirschvogel Automotive Group, the strategy consultancy McKinsey & Company Inc., Otto Spanner GmbH (specialising in pressed, stamped, and drawn parts as well as assemblies made of metal), Schnupp GmbH & Co. Hydraulik KG, and the medical consulting service TeleClinic.

Among the supporters of the project [8] were: the technology provider Carraro Group; the TUM spin-off CipLab with its focus on lithium-ion battery systems and forecasting methods for the real-time measurement of battery conditions; CONTACT Software GmbH, the start-up Continuity for the development and provision of digital control units; Flugplatz Jesenwang GmbH for the provision of a test site for vehicles; Green World Ventures with its network of farmers and raw material processing such as the Moringa tree; Hella KGaA Hueck & Co. with a focus on lighting technology and electronic components; the Mahle Group with its expertise in mobile solutions for the automotive industry; MWS Industrieholding GmbH, which specialises in aluminium casting; Piaggio Commercial Vehicles; Robert Bosch GmbH; Thyssenkrupp; voxeljet AG, which specialises in industrial 3D printing; WATTSTUNDE GmbH, which specialises in solar modules, special designs and custom products; the Wittenstein Group, which specialises in innovative drive technology; and ZG GmbH, which specialises in electrical and hybrid drive technology. Bayern Innovativ also supported the project [9].

The funding body was the Bavarian Research Foundation (Bayerische Forschungsstiftung) [10], which had funded the project with 923,000 euros over a period of 3 years. The total project volume amounted to 1.86 million euros.

For the explorative phase, the company also focused on Asian markets. This is because the vehicle structure of the Asian market is reflected in East Africa in particular. In order to meet the growth targets for the development of rural areas, a certain vehicle structure is also needed there. They therefore oriented themselves to the approved vehicle classes and deliberately looked for vehicle types that would do justice to the local competitive situation in Ghana. In Europe there is a lightweight class L7e [11], which was suitable for alternative drive technologies [12]. Countries such as China and India orient themselves on well-known standards. The de facto standardisation to be expected as a result also gives hope for a larger sales market with a view to the future international positioning of the aCar.

6.3.2 Phase II: Vehicle Concept

In the search for suitable vehicle criteria, local farmers were involved. In addition to Ghana, work was also carried out with the farmers on site in Nigeria, Kenya, and Tanzania. A trailer built by the research team itself was presented as a preliminary stage of the vehicle.

Rapid prototyping provided clarity about the functionality of the vehicle to be used and its design [13]. Using sketches, dialogue, and joint workshops, prototypes were developed and further refined. The contacts at local universities and the agricultural chair helped to establish contacts with farmers and to mediate. The votes were taken in English with the support of local contacts. Above all, it was necessary to establish a relationship of trust. An iterative process ensured that technical and utilisation issues could be clarified directly on site [14]. It turned out that the optimal loading area measures two rice bags lengthwise. The rice bag is the euro-pallet of Africa.

In mid-2014, the finished concept and the prototype were ready. In analogy to the scrum procedure in software development, a physical scrum was created iteratively. Again and again the farmers were able to understand what was being developed and why. If there were inquiries, these were explained. If functions were forgotten, they were added—physically as well as conceptually. The selection of the target group of farmers and research into living and working conditions contributed significantly to the acceptance of the vehicle. Through workshops in the format of design thinking and joint discussions, customers and the product could virtually find each other and shape the innovation process on site. The result is a vehicle concept based on the pick-up (Fig. 6.2).

However, the aCar differs from previous vehicle models in its multifunctional use of loading space and its versatility (Figs. 6.3, 6.4, 6.5, and 6.6).

Fig. 6.2 The aCar—Representation 1. (Courtesy of the Technical University of Munich/Mobility Systems Engineering)

Fig. 6.3 The aCar—Representation 2. (Courtesy of the Technical University of Munich/Mobility Systems Engineering)

Fig. 6.4 The aCar—Depiction 3. (Courtesy of the Technical University of Munich/Mobility Systems Engineering)

6.3.3 Phase III: Specification and Construction Phase

The decisive project phase started from January 2015 to December 2017. Here the specification of the target agreement, the updating of the concept in the specifications, the technical construction and manufacture of two test vehicles took place. In the meantime, the project team consisted of almost 100 students and 20 doctoral students.

Fig. 6.5 The aCar—Depiction 4. (Courtesy of the Technical University of Munich/Mobility Systems Engineering)

Fig. 6.6 The aCar—Depiction 5. (Courtesy of the Technical University of Munich/Mobility Systems Engineering)

The effort of the 3-year construction phase based on an extensive functional depth, oriented towards the user, can only be guessed at here: from the design of the vehicle and its applicability on site, the selection and use of non-vehicle-specific but already existing components, the advertising of and cooperation with suppliers, to individual tests based on components and the testing of the assembly, as well as the driving capability of the

Fig. 6.7 The aCar cockpit. (Courtesy of the Technical University of Munich/Mobility Systems Engineering)

vehicle. Within 3 years, the idea of a mobility concept for farmers in Ghana turned into a finished design for the aCar. A total of two aCars were produced in mid-2017. The aCar offered farmers above all simple handling in operation, as a glance into the cockpit reveals (Fig. 6.7).

6.3.4 Phase IV: Test Phase

Before the research team was able to take care of the test deployment on site, the vehicle prototype P1 was tested by the regulators in Germany. The P1 was then used by the farmers for a period of 3 weeks (Fig. 6.8). The construction of the P2 prototype started in summer 2017 [15].

The construction method and the use of solar energy (Fig. 6.9) are a further plus, as was shown in the test phase:

> "In July, scientists tested the first prototype of the aCar in Ghana. Not only do the road conditions differ greatly from German conditions with little asphalt, not always flat, but all the more sandy stretches, but also the climate with its high temperatures and humidity places different demands on an electric vehicle. The functionality of the components and the off-road capability of the prototype could be ideally tested in the Ashanti region around Kumasi, whose geographical conditions are representative for many African countries. The challenges of the circuit drives under local conditions and the bad road tests posed no problem for the vehicle and its electric drive train. The most important finding, however, was that the local people were very enthusiastic about the concept," says the Chair of Automotive Engineering at the Technical University of Munich.

Fig. 6.8 The aCar in action. (Courtesy of the Technical University of Munich/Mobility Systems Engineering)

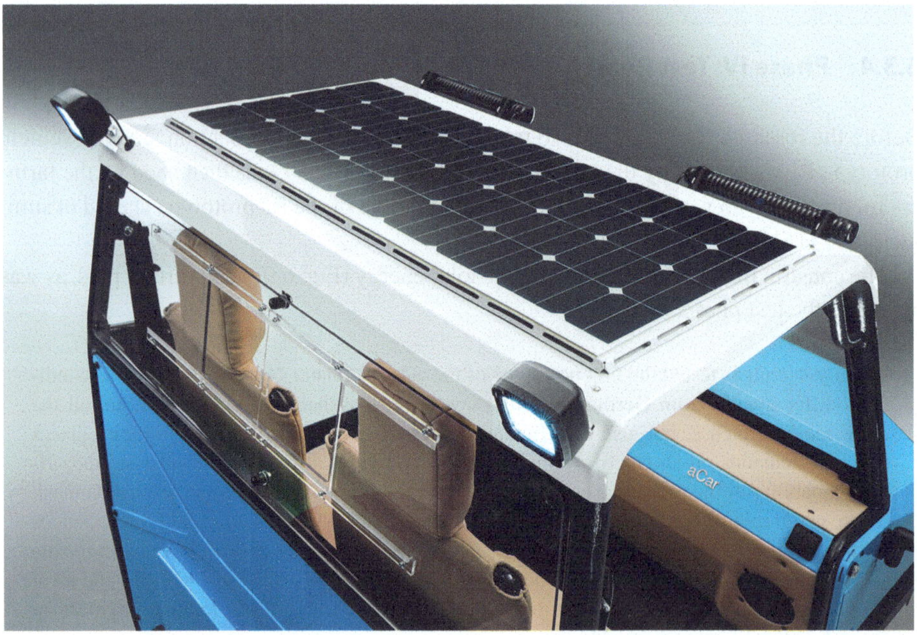

Fig. 6.9 The aCar—solar panel. (Courtesy of the Technical University of Munich/Mobility Systems Engineering)

The chairman of the university, the King of Ashanti, had driven the aCar himself, according to Sascha Koberstaedt. This resulted in a high level of attention for the project in local reporting. Importantly, there was an increased interest in the further course of action. According to Šoltés and Koberstaedt, the great response brought around the time to consider series production and the optimisation of the vehicle concept.

A central result of the trial phase was that the aCar, as part of the everyday lives and livelihood of farmers and their families, secured and did more than just serve the farmers' mobility needs. The aCar intervenes much earlier in the farmers' value chain—it functions as a charging station, helps with field cultivation, and is available to the power supply in the house as an auxiliary unit and as a light source. In future, the raw materials can be preserved on site and processed directly into flour at the farm. The intermediate product is offered on the market at a corresponding price. The protection of the goods is ensured during transport and delivery.

> **The Vehicle Becomes a Multifunctional, Value-Securing, and Value-Adding Space** The aCar gives its users a new value. It becomes part of the value chain of farmers and their families: it becomes a charging station, a power supply and auxiliary unit, light source and refrigerator. It is available as a minibus for transporting people and goods and as a mobile pharmacy and emergency service. The aCar promotes the handling of innovative drive and mobility applications and offers a training perspective.

6.3.5 Phase V: EVUM Launch

The start-up EVUM Motors GmbH was founded in 2017. The founding fathers are Martin Šoltés and Sascha Koberstaedt. Since its foundation, EVUM Motors GmbH (hereinafter referred to as EVUM Motors) has been planning to build ten vehicles at series level by the end of 2018. Five of the vehicles are planned for use in Africa, in particular Ghana. A target size of 1000 aCars is scheduled for the end of 2019.

As things stand at present, the first financing round for the start-up in 2018 will be set up. How scalability can be managed by applying network effects is explained in the following chapter.

6.4 Sustainable Mobility for Emerging Countries

6.4.1 Let's Ideate the Journey from Ghana to the World

After the successful initiation phase, DoktorB has been elaborating on the scalability options as EVUM Motors is focusing on the suitability of the vehicle for series production. The first experiences in Ghana continue to play an important role and provide initial insights into the production, construction, use and distribution of the vehicle. Figure 6.10 provides an overview of the main planning steps.

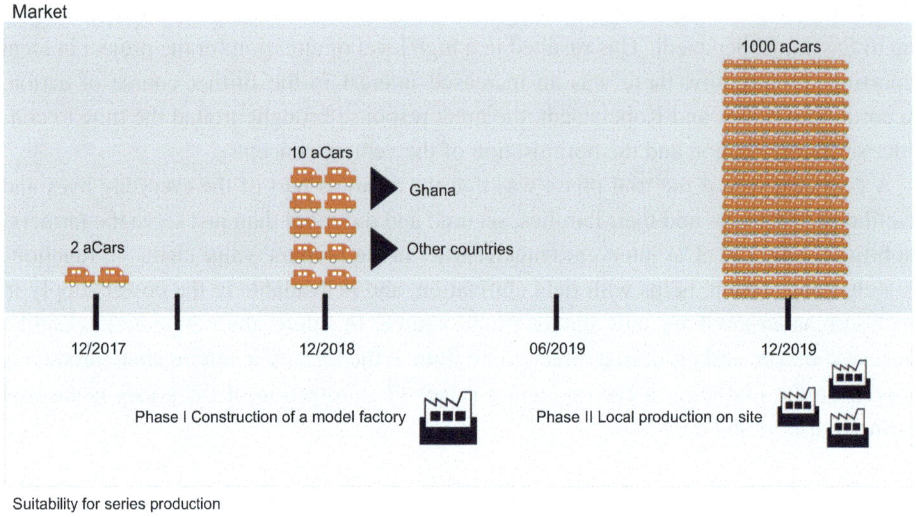

Fig. 6.10 Scalability plan for EVUM Motors Ltd.: from production to market growth

The colleagues continue the project on series suitability as they started it in the first sketches. In order to achieve the greatest possible benefit for the farmers, it is necessary to use existing technology, concentrate on the prefabrication of components, and enable on-site assembly with known and trained personnel. The construction of a model factory in Germany is the next milestone. A look at the full cost calculation shows that the manufacturing and assembly costs are still too high. The conception of a vehicle as a construction kit should contribute to a cost reduction, especially with regard to delivery and assembly costs in the target country.

Subsequently, local production and assembly are planned directly on site. EVUM Motors' business model could be as follows:

- Selection and qualification of component manufacturers and suppliers
- Qualification of assembly partners
- Creation of a franchise concept for production and assembly on site
- Selection of franchisees by EVUM Motors
- Assembling, distribution, and maintenance by franchisee
- Research and development remains with EVUM Motors
- Depending on the merger, local farmers offer themselves as buyers and users of the vehicles. In the Verbund, as a cooperative formation, the farmers then leverage the traditional intermediaries such as the truck operator.

A successful franchise concept starts with the selection of local partners and their support in setting up the organisational structures, such as instruction in the construction plan and assembly steps. Qualification of employees on site through train-the-trainer training courses sets the course for acceptance of the aCar. These and other measures, such as mobility services and other product-related services, also ensure that the local population is directly involved and becomes part of the EVUM Motors team! This in turn brings us a little closer to our goal of helping everyone on site to become more mobile.

In order to achieve the planned sales volumes of aCars, further markets outside of Africa are needed. The target figures according to Fig. 6.10 can be achieved by using the aCar in those countries which, like Ghana, have a comparable level of income and a similar dependence on intermediaries with regard to storage and distribution of raw materials and food. We are thinking here of countries such as Nepal and Bangladesh. The Human Development Index [16] provides assistance here.

6.4.2 An Ecosystem for Sustainable Mobility

Without prejudging too much the organisational understructure for EVUM Motors, the ecosystem for the aCar will be characterised by:

- A business model geared towards cooperation and the common good
- A manageable vertical range of manufacture and a flexible supplier structure
- A versatile application despite geographical and climatic topologies
- A typification of the production and sales partners to be involved, as well as
- A successful implementation of a Mobility-as-a-Service approach to weave the aCar into local sales and distribution structures

Depending on the country of deployment, the latter will determine whether a cooperative buys the vehicle and offers it to its members or whether a mobility provider establishes itself as an intermediary. The stringent pursuit of modular vehicle construction influences the network design; the acquisition of local partners creates local added-value, low-cost production, and jobs. These are created in connection with component production and assembly, amongst other things.

EVUM Motors is of limited interest to OEMs. In the small vehicle segment, EVUM Motors could be considered as a sales partner for new cars in growth markets such as Africa. The agricultural sector is an important industrial segment, as the project in Ghana has shown. The use of the aCar is interesting for those branches of industry and customer structures which are characterised by small-sized products as well as cooling and heating chains.

Looking at India, the aCar is a good addition to the Renault Kwid as a multi-purpose vehicle, says Sascha Koberstaedt. Overall, however, OEMs are too sluggish according to their own experience; the internal speed to initiate structural changes, for example to supplement the vehicle fleet with a model such as the aCar, is too slow.

6.4.3 Securing the Survival Space in the Developing Country

As things stand today, one can speculate about the development of the price structure that aCar buyers and users are facing. It can be assumed that several families will purchase an aCar in the future and share its use. Otherwise, there is a danger that the current supremacy of truck operators will only increase if they secure aCars early and unnecessarily drive up the usage price.

It therefore needs political support on the ground, as well as partners who can help with industrialisation and licensing. The response of the farmers during the trials has shown that complex, innovative projects with the necessary traction and which consider the inner-drive of the project participants and project promoters are very practical and feasible.

Ghana itself is no less distinguished by its claim to establish an automotive sector—according to the observations and impressions of the project team. Quite a few components and dependencies need to be taken into account in order to set up local parts production, distribution, and usage structures on site and to assemble the vehicle.

Innovation projects such as the TUM help countries such as Ghana to a new status. The decisive factor will be the successful implementation of series production within a tolerable time frame. The project initiators and project members have proven that the claim is feasible: to launch the most sustainable form of development aid!

▶ **Positioning on New Terrain** Beyond industrial and consumer-oriented functionalities, altruistic innovation projects in developing countries offer the automotive and supplier sector the opportunity to question the role of the car and to assign new tasks to both the vehicle and the industry itself.

6.5 DoktorB's Checklist

Important Basic Considerations

- Do you see the car in your everyday life as a survival space?
- How do you define "survival space" at all?
- Where do you think the vehicle will impact the farmers' daily lives the most?
- How much time do you think farmers and families will spend in the car?
- Where could the aCar be useful to you?
- Which functionalities would you still trust in the vehicle?
- In your opinion, which risks are reduced or even eliminated by a multi-functional vehicle?
- What opportunities does the research team's approach offer your own projects?

- At what price would you offer the vehicle?
- What does the dialogue with users look like for you?
- When was the last time you spoke directly to the users of your solutions?
- When was the last time you asked your customers for feedback and input on a product idea?
- Public administrations, especially in rural areas, are looking for cooperation with industrial and supporting partners—what does your partner landscape look like for your project?
- What role(s) and tasks could your organisation take on based on the above description?
- How does your organisation, community, or city deal with testing new "hardware"?
- When did your organisation test a multi-functional product beyond its previous use? How quickly was multi-functionality discarded as reverie?
- How long does the innovation process take for you—from the idea to prototype testing and production readiness?
- How much have innovations changed your collaboration with customers and partners, suppliers and public administrations?

References

1. TUM (2017) aCar mobility – rural mobility in developing countries. http://www.acar.tum.de/index.php?id=5. Accessed 17 Feb 2018
2. Website of Prof. Lienkamp, Chair of Automotive Engineering at the Technical University Munich (TUM). https://www.ftm.mw.tum.de/index.php?id=316. Accessed 17 Feb 2018
3. Website of EVUM Motors (2018). Accessed 16 Dec 2017. http://www.evum-motors.com/. Accessed 17 Feb 2018
4. Wikipedia (2017) Ghana. https://de.wikipedia.org/wiki/Ghana. Accessed 17 Feb 2018
5. Institute of Food and Nutrition Sciences, ETH Zurich (2009) The Manioc Plant. http://www.cassava.ch/maniok/pflanze.htm. Accessed 17 Feb 2018
6. TUM (2018) aCar mobility – presentation. https://www.acar.tum.de/index.php?id=5. Accessed 17 Feb 2018
7. TUM (2018) aCar mobility – industrial partner. http://www.acar.tum.de/index.php?id=92. Accessed 17 Feb 2018
8. TUM (2018) aCar mobility – supporter. http://www.acar.tum.de/index.php?id=93. Accessed 17 Feb 2018
9. Bayern Innovativ (2017) Cluster automotive – aCar mobility – a mobility concept for Africa. http://www.bayern-innovativ.de/e38aec1d-0df0-c16c-5ee3-25f91ad14589?PP=4759a759-edc8-45ca-9523-a359acb70da0. Accessed 17 Feb 2018
10. TUM (2017) aCar mobility – project funding. http://www.acar.tum.de/index.php?id=27. Accessed 17 Feb 2018
11. Wikipedia (2017) EC vehicle class. https://de.wikipedia.org/wiki/EG-Fahrzeugklasse#Klasse_L. Accessed 17 Feb 2018

12. Simon P (2009) Development of a light metal space frame structure for vehicles with alternative drive concepts. Lightweight Des 2:26–33
13. TUM (2018) aCar mobility – mobility data and usage scenarios. http://www.acar.tum.de/index.php?id=22. Accessed 17 Feb 2018
14. TUM (2018) aCar mobility – concept development and design tailored to needs. http://www.acar.tum.de/index.php?id=24. Accessed 17 Feb 2018
15. TUM (2018) aCar mobility – vehicle development. http://www.acar.tum.de/index.php?id=20. Accessed 17 Mar 2018
16. Human Development Index (2018) website. http://hdr.undp.org/en/content/human-development-index-hdi. Accessed 18 May 2018

Sustainability on x-Number of Wheels

7

DoktorB Barbara Flügge

Abstract

Many transport providers would like to meet the demand for sustainable mobility. In the current climate, willingness to get involved in Mobility as a Service (MaaS) is just as crucial as technological innovations, and companies need to work together with competitors and business partners. The Green Class example, discussed in this chapter, demonstrates how MaaS can work, but also that implementation cannot (yet) be secured for everyone; financial and infrastructural resources and openness in political bodies to adopt innovative concepts can be difficult to achieve.

In thinking about the car in conjunction with living space and intermodal offerings, two questions arise:

- How emotional is the car?
- For what reasons does travel via the car occur, even when resulting in traffic jams or city driving?

In small and medium-sized communities, those in charge prioritise an offering's survival in rural areas, beating opposition, and becoming attractive to the existing popular and new popular groups.

D. B. Flügge (✉)
digital value creators (DVC), Bottighofen, Switzerland
e-mail: barbara.fluegge@dvcconsult.com

B. Flügge (ed.), *Smart Mobility in Practice: Good Car, Bad Car, No Car – Is this the End of Nurturing Our Mobility DNA?*,
https://doi.org/10.1007/978-3-658-43171-6_7

For larger municipalities, the challenge comes from compliance with sustainability goals. In these cases, regulatory requirements and targets work alongside the "parking space key". The parking space key denotes how many parking spaces are made available for one building. For example, the current 0.8 value specifies that eight parking spots would be offered to ten units (a figure of 80%). In Zurich, we have observed a decline to 0.4, or even 0.0 in residential complexes, as focus shifts towards an offering of one community car and/or car sharing for the neighbourhood and the obligation to use public transport.

This chapter discusses the question of ideal, sustainable travel and the acceptance of alternative means of transport, reflecting on the Green Class Model Project [1].

7.1 Destination Station: Intermodality in Practice

With the focus on sustainable mobility, the importance of the car will have to be considered from two different points of view: (a) How sustainable is the car's life cycle from origin, through use, and up to disposal or reuse and (b) How sustainable is the intermodal chain when using a vehicle?

Concerning our first question, the study book "Cradle to Cradle: Remarking the Way We Make Things", with an introduction by Prof. Braungart [2], is great in the discussion of the Circular Economy movement [3]. Another down-to-earth book that DoktorB recently published simulates the benefits of reitering the life cycle of urban areas into urban farming. We all agree that a comprehensive discussion of the life cycle of vehicles is needed. Alongside cars, this covers buses, lorries, delivery vehicles, vans, and other relevant transportation means. In this chapter, we focus on the second question (namely b) and ask to what extent the matter of sustainability influences the functional set-up and size of the car.

Let's dive into this field. If we are not travelling by road via only one vehicle, but are changing traffic modes, we are intermodally dependent on providers and operators and their sustainable value proposition. Electromobility is used in vehicles, as well as by rail and bicycle rental companies. In addition to the car, city, and country buses should also be considered. An isolated mobility supply of exclusively rail and cars would be a disadvantage. On an intermodal route with a single flight segment, the energy input takes a considerable amount of time compared to a train. The railway is more sustainable than the combustion engine. If you book a ticket today on the SBB.ch website, you will be eligible for the CO_2 value and also receive an indication of the fuel consumption saved in comparison with a conventional vehicle. A general statement can be found here [4].

So where can you benefit from intermodal mobility, not only compromising but taking actual advantages? How about that? The consumer's decision is based on the offered combination. The administrative effort is lowered to a minimum. However, in case we feel trapped in a lock-in effect, we would like to have decided otherwise—if we could. We would need to sort out individual and separate alternative travel options that suit us.

The stakeholders who created the so-called Green Class offering are Swiss Railways (SBB), PubliBike, and BMW. They joined forces in the Green Class project to promote sustainable mobility. The success of this project should be measured by the following criteria:

(a) At peak times and when there is expected overcrowding on trains, the Green Class Ticket should cushion the use of the train and steer travellers towards the car.
(b) Missing sections of track, lack of range and lack of flexibility of the rail offer will be completed by the offer of a car.

Rail transport alone has no far-reaching mobility. In this sense, achieving a far-reaching service offer means including a combination of car, bike, and train services to compensate for missing track sections, to offer agility in small spaces and to include the missing last mile for delivery at home or at work. From the point of view of an automobile manufacturer, Green Class complements the portfolio and helps BMW reach out to new customer groups.

From the three service providers' (SBB, PubliBike, and BMW) points of view, the advantages of such a service lie in their complementary offers. When combined into a whole, the car can be used where the train cannot travel or is too inflexible. The same applies to the bicycle. Conversely, the train can be used on routes that are no longer fun for a motorist today anyway, as through traffic jams and construction sites the daily routine and the mood can be madly whirled up. Through the combination of rail, bicycle, and car, competitive advantages have been worked out to the benefit the three stakeholders, too.

7.2 On the Road in the Green Class

BMW, PubliBike, and SBB initiated the Green Class [1] project with the aim of attracting 100 participants from the Swiss population to test the Green Class. On board were the ETH University of Zurich [5] as partners and Alpiq [6] and EWZ [7] as service providers. Green Class is a mobility offer consisting of three means of transport or their services. For 12 months, the Green Class user was given:
A vehicle, the BMW i3, with mileage limitation
The unrestricted use of SBB in 1st class,
The unrestricted use of the supplier's bicycles *PubliBike* [8]
The aim of the initiators was to facilitate the change from one means of transport to another, and to reduce infrastructural disadvantages in the communities by alternative means of transport. So, for example, someone uses the car to drive to the next station parking lot and changes to the train. In the city, the passenger uses this for the last mile from the train station to the office or shop.

The Green Class

The Green Class mobility offer from SBB and BMW consists of free use of a BMW i3, the GA travelcard (GA stands for General Abonnement) gives you free rides on SBB train services and on other public services e.g. ferries) for 1st class, and free use of PubliBike bicycles for a period of 1 year. Participants pay a participation fee of 10,000 Swiss francs.

The Green Class project was launched throughout Switzerland. The aim was to create a more suitable mobility offer for people living on the outskirts of a city, of which there are many communities. Their challenge is to meet the expectations of commuters and the needs of the population. Often there are small waiting areas and long waiting times for the next train or bus. For a concrete intermodal offer, the docking stations to other means of transport are missing in many communities.

Many responded to the call to participate in the Green Class project. The offer was then implemented in 2016 with 150 participants. The participants were selected by means of a questionnaire on their willingness to implement the parking space or connection for electromobility at home, information on commuting and travel behaviour, and their willingness to have the driving and usage behaviour of the means of transport on offer documented and evaluated. Every Green Class participant used the Green Class app for tracking. A project team from ETH Zurich carried out the evaluation of the usage behaviour and the provided car's used mileage range, as well as a survey on the product mobility offered by SBB and BMW.

An evaluation of the applications received and accepted would be worthwhile, particularly regarding;

- The proportion of urban and rural applicants
- The regional distribution of applicants according to German-speaking Switzerland, French-speaking Switzerland, and Italian-speaking districts
- The demographic structure

Concerning the usage behaviour, we look forward to a long-time study report discussing:

- The redistribution of kilometres travelled compared with the rail kilometres or routes newly and regularly used by Green Class
- The perception of a mixed transport offer from the point of view of railway drivers and from the point of view of motorists
- The consistency of the Green Class offer beyond the trial phase
- The associated change in the behaviour of car drivers
- The use of electromobility

Turning Away from the Car as a Habitat?

A new understanding of mobility is needed to continue the Green Class project. On the part of motorists, understanding with regard to using a second means of transport and, despite traffic jams and the need for parking, relying on a combination of several means of transport to get ahead, is needed. Without this understanding, no further test drivers and users would be found. Another parameter is the price for the Green Class ticket. In order to deal with the current annual amount, it is important to understand one's own previous mobility costs. In considering this, we carried out a comparison of effort and budget (cf. [9], p. 218).

Many drivers do not know their mobility costs. The situation is quite different for rail passengers. In order to attract further user groups for Green Class, transparency about costs, efforts, and benefits is needed for and amongst car drivers. Only then does the contribution of 10,000 Swiss francs become justifiable for those interested. The unique selling proposition of Green Class will hinge on whether it is able to trigger emotion within a few seconds to transforms the service into an attractive one.

Driving is freedom. Drivers therefore find it more difficult to give up the perceived flexibility of having a continuously available car. The own vehicle is neither dependent on given departure and arrival times, nor on externally caused changes in the timetable or cancellations of the public transport system and its transport means.

In addition to the symbol of freedom, we perceive the vehicle as a safe place. We control access and entry. Analogous to our apartment or house, we reserve the right to share this place. Whether the comfort of one driver meets the demands of another is a matter of taste. Despite the existing comfort on European trains, we perceive the car as the actual comfort zone. After all, we were right to consciously choose and furnish the living space of the car, the interior, the functional equipment and the sound system. The equipment that did not fit into the budget is quickly forgotten. And we are happy about the equipment that we could and wanted to afford.

The Green Class offer, consisting of a train card, vehicle, and an access code to PubliBike, puts car fans and their desire for unlimited range and flexibility to the test at any time—they are now dealing with electric mobility. With regard to the range of the car, this depends on the average distances of the Green Class user and how he divides his route planning between rail and car. We remember that Green Class is using an electric car with the BMW i3.

If the daily range remains below the maximum range of the BMW i3 when the battery is fully charged, Green Class users will not have to make any unnecessary efforts. The current range of the BMW i3 is 160 km [10].

What If

In the above remarks, we have referred to a number of variables that determine the wide acceptance of Green Class. Let's summarise the essential points; an acceptance formula resonates if we encounter:

- Planning expenditure: Planning expenditure for routes, parking space booking and means of transport occupancy must not increase.

- Digital support: Intermodal planning and booking functions in the Green Class app for PubliBike, car parking at central stations, parking spaces in other preferred locations.
- Time savings: Find, select, reserve, and make available charging points, speed up charging processes for electric cars, increase range for electric cars.
- Proof of costs: Intermodal mobility demonstrably pays off compared to previous expenditures.
- Range: Improve the range for electric cars. It also requires a corresponding number of loading points as well as the possibility of using loading points outside opening hours at loading locations such as in the parking garages of shopping centres or private providers.

We are well aware that the parameters mentioned with respect to electric cars, in particular, require a solution not only in connection with Green Class. Technical solutions as well as economic and administrative measures are fundamentally required here in order to achieve a higher conversion rate to electric vehicles.

For those who are in favour of a Green class offer but who already own a vehicle, the question also arises as to how they should handle their own vehicle. For example, in a two-car household, while the older model has long since paid for itself, a new model has just been added at a favourable price via a leasing contract as a second car. Others who have not yet had their own vehicle and would like to use Green Class are confronted with the search for a parking space. Here, a suitable booking system for on the go, via the Green Class app, can help. It's different at home. If you don't have your own parking space, it gets tricky in some places. Here, Carsharing would be the obvious choice within the household or between neighbours. This would then require a distinct advertising strategy.

Here again, attention should be paid to the charging infrastructure (see also Sect. 9.2); the success of electromobility depends to a large extent on an intelligent infrastructure that is made available. "Park and Rail" would be used much more often if loading points were available at stations. For the Green Class operator there is a desired side effect: as Green Class users, we learn how to handle charging the electric car and how tricky the arrangement of the charging station can be compared to the length of the charging cable. What does the business model for the charging infrastructure look like at all? And what support do we need in the Digital Front Office (see [9], p. 69)? A seamless outdoor/indoor navigation could already book us the suitable parking place within reach of the used railway track!

7.3 With the Green Class from St. Gallen to Oslo

The question is, how could SBB and partners scale the Green Class? Would it be possible to travel from St. Gallen to Oslo with the Green Class? We have considered the following scenarios:

1. Sustainable Company-Wide Mobility

Green Class could be integrated into existing ridesharing models. The company SAP® makes the TwoGo [11] offer available to employees for this purpose. With this or similar offers, Green Class users would also have conveniently located and easily accessible parking spaces on the company's own premises. The demand for sustainable mobility is met by the use of electric vehicles—passengers learn to handle these, and also to trust the charging capacity and availability of charging points. Rituals can develop well and lead to a change in behaviour if they are experienced regularly and in a group, such as when commuting to work or to joint venues for events and meetings.

2. Mobility as a Service

Green Class becomes part of a city's Mobility as a Service offering. Cars and trains have taken a stronger place in existing public transport services for urban areas. The ideal solution would be to implement the European Union directive, which provides for a single local transport ticket and relieves users of the administrative burden. Depending on the purpose, space requirement and urgency, the car, or alternatively the bicycle, the tram, the rail or ridesharing, is used. A community pooling system ensures good capacity utilisation if residents opt for car sharing.

3. Rediscover User Groups

How do MaaS providers attract new user groups? Green Class, for example, is a pioneer in the field of car rentals, which is attracting more and more interest from those who have an affinity for cars and have hired a car from friends. This opens up customer groups who have not even thought about using a car. The reasons for the use of the system are illustrated by the analysis of primary, secondary, and tertiary interests in some examples in ◉ Fig. 7.1.

These reasons can be manifold. For example, the purchase of a car may be too expensive, the search for a parking space too time-consuming, or the car may be regarded as harmful to the environment if it has previous technologies. Other reasons lie in the fact that the acquisition of an own vehicle is not worthwhile and possession is not up-to-date and/or is superfluous. If it functions, borrowing, parking, and loading, nothing stands in the way of newly gained freedom. If the driver's license is missing, autonomous vehicles will be available in the future!

4. All a Question of Class?

The applicants for the Green Class project had deliberately opted for a premium offer. What would the Green Class business model for second class look like? And what contribution would SBB, BMW, and PubliBike have to make to persuade many non-car drivers and railway avoiders? Could one imagine a cooperation with the peripheral communities, which today are little connected? Additionally, what would a functioning charging

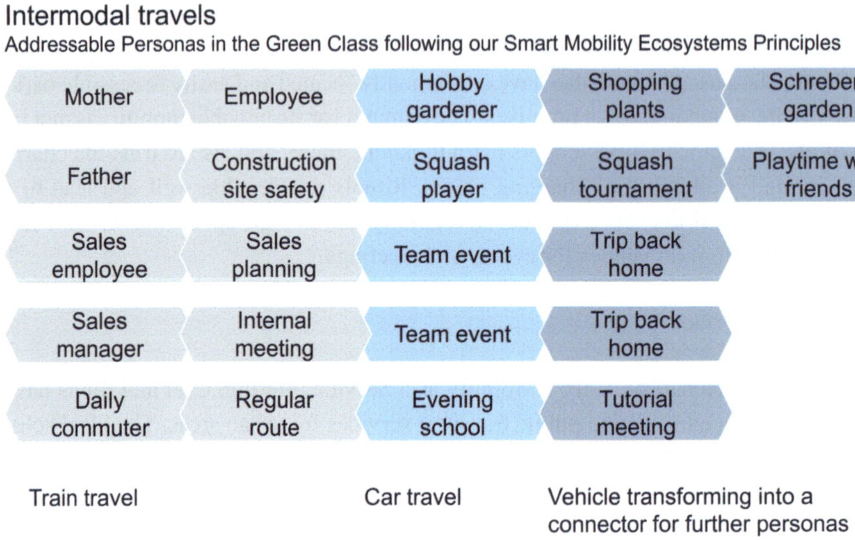

Fig. 7.1 Addressable personas in the Green Class

infrastructure look like with current user growth, and that would also do justice to the targeted growth? From our point of view, these are the essential questions that need answering to make it possible to offer Green Class even beyond the test phase. The latest information comes from the Geneva Motor Show:

> The continuation of 'SBB Green Class' was announced in the presence of Federal Councillor Guy Parmelin at the BMW Group stand on the Geneva Palexpo site. 'SBB Green Class' is a flexible and cross-modal offer for sustainable mobility. During the pilot project it included a general SBB subscription for 1st class travel (SwissPass), which also included public transport services, as well as the use of a purely electrically powered BMW i3 with BMW i Wallbox for charging. In addition, customers were able to use the services of local car and bike sharing providers as part of SBB Green Class. [12]

With the evaluation of usage figures and the consideration of inputs by users, the above-mentioned considerations will become more concrete and new scenarios will be added.

The fourth scenario is also accompanied by the question of the equipment of the vehicles. In the context of autonomous driving, we are already discussing different levels of equipment and the associated surcharges that occur at the time of booking. Is this actually applied in a geographically dense area? Here, too, suitable simulation models and the creation of impact diagrams are required to design the optimum mix of vehicles and locations according to the site profile and topology. Equipment and comfort are also essential for motorists, who need to be served when changing from car to train. Those who want to take over a car also expect attractive waiting points and lounges. Today, waiting rooms are rare and sparsely equipped and often far away from the actual departure points, and even further from the nearest road where a Green Class car is handed over.

The catalogue of requirements for the establishment of Green Class and similar projects is comprehensive. We strongly recommend the following points:

- As described above, simulations and impact diagrams are needed for the acceptance and expansion of the Green Class innovation programme.
- Creation of infrastructural preliminary work and a foundation by public and private companies for the acceptance of electric mobility as a real alternative to the combustion engine.
- The installation for the charging box (so-called wallbox) at the user's home must be solved without effort for the user.
- Existing platforms from e-ticketing and car sharing to digitally usable timetables must be used. Only then can one find and maintain one's own positions in local and international competition. This applies both to the Green Class project and to the individual undertakings concerned.

If I think back to the questionnaire that was presented to the applicants for the first phase of Green Class, it helped to select the Green Class applicant. A similar selection procedure will also be unavoidable for further expansion stages, particularly with regard to electromobility.

MaaS Is Taking You Anywhere in the World

A cross-national scaling of Green Class is not only conceivable, but also meets the requirements of a genuine Mobility-as-a-Service offer. Imagine what a trip from St. Gallen to Oslo could look like with Mobility-as-a-Service and Green Class!

7.4 DoktorB's Checklist

Important Basic Considerations

- Why would you decide on a mobility mix à la Green Class?
- Do you know your mobility costs?
- As a railway supplier or car manufacturer, would you support a project such as Green Class in your country or federal state?
- Think of your community: Do you spontaneously think of citizens and community employees who would be enthusiastic about a MaaS offer?
- What support do you need as a provider of e-bikes and/or cargo bikes?
- What measures are needed to increase the use of bicycles—in connection with a car and the train? Does it need a bicycle highway like in Copenhagen?
- What are the three top functions that a digital companion would have to offer you in order to actually use a flexible range of rail, car, and bicycle services?
- Name your possible cooperation partners for a project in the sense of intermodal mobility!

References

1. SBB (2016) Pilot 2016: SBB Green Class E-Car. https://www.sbb.ch/en/travelcards-and-tickets/railpasses/greenclass/pilot-projects/e-car.html. Accessed 18 May 2018
2. Ellen MacArthur Foundation (2017) What is a circular economy. https://www.ellenmacarthur-foundation.org/circular-economy. Accessed 18 May 2018
3. Braungart M, McDonough W (2003) Simply intelligent production. Berliner Taschenbuch Verlag, Berlin. ISBN 978-3-8333-0183-4
4. SBB (2018) Travelling by train means travelling in an environmentally friendly way. https://www.sbb.ch/de/geschaeftsreisen/mobilitaetsberatung/oekologisch-reisen.htmlm. Accessed 18 May 2018
5. SBB (2017) Exciting findings from the mobility data collected by SBB Green Class. https://dima.sbb.ch/unterwegs/artikel/62834/spannende-erkenntnisse-aus-den-gesammelten-mobilitaetsdaten-von-sbb-green-class. Accessed 18 May 2018
6. Alpiq (2017) Company website. http://www.alpiq.com/portal/en/. Accessed 18 May 2018
7. EWZ (2017) Company website. https://www.ewz.ch/de/private.html. Accessed 18 May 2018
8. Publibike (2017) Company website. https://www.publibike.ch/de/publibike/. Accessed 18 May 2018
9. Flügge B (ed) (2016) Smart mobility – trends, concepts, best practices for intelligent mobility. Springer Vieweg, Wiesbaden
10. BMW (2017) BMW i3 website. https://www.bmw.de/de/neufahrzeuge/bmw-i/i3/2017/reichweite-laden-effizienz.html?uncert. Accessed 18 May 2018
11. TwoGo (2017) Company website. https://www.twogo.com/. Accessed 18 May 2018
12. AUTONEWS.CH (2018) BMW Switzerland and SBB continue. SBB Green Class. http://www.autonews.ch/?p=19003. Accessed 18 May 2018

Innovative on the Road: Who Invents and Who Follows?

8

DoktorB Barbara Flügge

Abstract

The basis of successful innovation consists of the ability to act, legitimation and productive application. Being successful requires assertiveness and marketing skills—especially in organisational networks and in the internal competition of established companies. For innovative mobility concepts such as autonomous driving, for example, newcomers and suppliers to the automotive industry are increasingly distinguishing themselves. Those who believe in innovation set off, regardless of whether their own position is called into question or not. Among the pioneers are the Styrian region in Austria with Alp.Lab, the ENTOURAGE research project, and SAP® with its decision to promote sustainable mobility beyond company boundaries.

Innovation advocates play as having an essential role here. Advocate sounds pugnacious. Being successful requires assertiveness and marketing skills—especially in organisational networks and in the internal competition of established companies. The same is true of those who have difficulty in the automotive industry. Innovation in the mobility segment requires organisational willingness to change. Irrespective of whether it is a question of technological, data-based, or other forms of innovation, the top management of a company should determine the willingness of its organisation to change on the basis of a few selected evaluation criteria ([1], p. 207), namely:

D. B. Flügge (✉)
digital value creators (DVC), Bottighofen, Switzerland
e-mail: barbara.fluegge@dvcconsult.com

B. Flügge (ed.), *Smart Mobility in Practice: Good Car, Bad Car, No Car – Is this the End of Nurturing Our Mobility DNA?*,
https://doi.org/10.1007/978-3-658-43171-6_8

- Getting to know the transformation maturity of your organisation and organisational footprint with respect to innovation,
- Assessing the reach of innovation, including the degree of complexity, expectations from the management and the markets,
- Determining to what degree the observed/desired innovation will influence and impact your workforce,
- Implementing an innovation-targeted procurement process and testing it for at least 30 days

The oil of the twenty-first century—data—is often the engine of innovation projects. In order to be able to use data and how it is handled in a profitable manner, automotive companies started with complementary services: entertainment, information about the nearest gas station, directions to nearby banks, or the indication of the weather to be expected for a pre-booked golf tournament. While other industries have already committed themselves to the concept of service engineering and IT-supported service innovation, a range of services in the entertainment or navigation system in the car has so far been sufficient. The first approaches to a Mobility-as-a-Service (MaaS) oriented offer have been manifested by DoktorB in the context of her business projects with Magna Steyr, Mercedes-Benz and university programmes. Other MaaS initiatives have been launched by BMW with DriveNow [2], Daimler, and the MOIA project [3] launched by Volkswagen in 2017. The consumers' demand for mobility at all times demands a radical rethink. The focus is not on the vehicle itself and vehicle-related features, but on offers fed into the vehicle, the mobile terminal and other devices and made possible by modern technology. And these should cover far more than vehicle-related services.

In the past, creative solutions have been sought to expand the *vehicle compartment,* and what is sought after, in order to offer as much comfort as possible in the smallest amount of space. Today this is one of the hotspots! Analogous to physical space, the data room is about design, availability, technical feasibility and cost: benefit ratio. In contrast to the traditional questions, however, the number of players is no longer limited to a few. These previous divisions have dissolved as far as possible, and players now find themselves confronted with questioning their previously valid supremacy and breaking away from it.

Hotspot Data Space

The data space becomes the desired hotspot for new business models and digital platform operators and ecosystems—in the vehicle and outside. Attributes such as design, availability, technical feasibility, and a comprehensible cost: benefit ratio are on the wish list of creative minds and innovators. Furthermore, intelligently linked mobility data complete the profile of your consumers. The tradability of data and the price offered to us will both influence our choice of vehicle in the future.

The diversity of technology providers must be sorted, and it must be determined which provider becomes a critical resource and which becomes the kingmaker. In game theory, the player who is not able to win is capable of determining who from the remaining will win. This asks for a willingness to transform entirely, from management towards organisational, production, and service levels. The company management sets the pace here. Let's take an automobile company named ABC. For a company like ABC that takes up the challenge and disruptively questions its own segment and its own position, these questions trigger the willingness and legitimacy to innovate. The interesting innovation projects are often relocated abroad or to top secret project locations or board units and special staff positions. This is where the futurologists of the company and its partners can be found. It can be assumed that test and market entry planning, in particular for new technologies, are well considered and prioritised to meet the top markets and geographies. Specific features make drivers ready to consider autonomous driving and prepare them mentally. German car brands focus on customer acquisition in the upper price segment. The recent decisions by Mercedes-Benz demonstrate the consequences of this focus. Customers take what's offered. Customers and local constituents are not consciously prepared to learn about innovation projects, nor are they actively involved, for example, through large-scale test rooms and participatory activities.

The Drive Me project by Volvo and the Smart Mobility programme launched by Ford are different. Ford's programme comprises of 25 innovation projects carried out in selected countries ([1], p. 109). In regard to Drive Me, Volvo started recruiting up to 100 families and individual volunteers for test drives in autonomous vehicles. The previous target number of 100 vehicles was postponed due to technical and logistical delays. Volvo's plan to involve people like you and I in the autonomous driving project has been criticised and called naïve. Volvo's announcement also led to a discussion on the appropriate timing to kick start autonomous driving tests in communities. The appropriateness of what is being tested and by whom has been criticised, too. Volvo's test announcement with 100 vehicles was even described as stupid in some places. One OEM told me that the estimated cost of one autonomous-enhanced vehicle lies at 200,000–300,000 euros per vehicle.

I ask myself in the face of this malice if the inventors of the telephone, satellites, aeroplanes, medicine, the power generator and other great inventions would have followed up their dreams if they had known about the omnipresent appetite for news, critics, and the overwhelming, sometimes incomprehensible, interest in failure? Would they ever had done something different? Maybe inventions would not have been invented at all? We know from great inventors that perseverance is a key characteristic to withstand side distractions and that is the audacious mind that overcomes obstacles. In DoktorB's course, the ABC Business Opportunity Design, we teach exactly this, overcoming obstacles for the inventors of the twenty-first century.

Regardless of innovation diversity within one industry segment, whether pre-announced or actually accomplished, brand addicted buyers are loyal buyers. It is us, the brand loyal buyers, who guarantee the continued success of brand car manufacturers. It is brand loyalty that carries on the use of proven and conventional technologies. Even without great innovations, service offerings with a glimpse of technology and market launches can be managed in such a way that we, as brand-conscious buyers, remain loyal to the company. The market can thus be skimmed off with a conservative or expanded, yet narrowly defined, range of services. This may sound critical, but it is a common and recognised feature of the free market economy.

The transformation of the automotive sector into a service and technology sector requires market participants to be aware of the corporate mission as employers. This is about safeguarding jobs and the economic viability of entire communities and regions. The latter element includes innovation leadership and the degree to which an innovation has a positive impact on the workforce, i.e. the ability to build competencies and talent and to ensure healthy economic and environmentally sustainable growth locally and within corporate structures. In the automotive industry, the network of locations, production and assembly sites, suppliers, assembly operators, and many other elements, has been carefully built and standardised. Everyone following the rulebook is perfectly performing their role, like the components in a Swiss clock. The book of rules is essential to keeping the OEM and car manufacturing business alive. It would be a massive transformation change—if not at least one successful disruptor like Tesla—that requires a master plan and suitable simulation mechanisms to prepare the existing network for innovation boosters, as well as internal organisational and operational changes. Disrupting conventional drive systems, and introducing electric, hydrogen-powered and automated drive systems, penetrates the DNA of the automotive sector and turns the mechanisms of vehicle production that have been established and proven for decades upside down, as well as hard-wired supplier and cooperation partnerships.

The traditional companies are contrasted by service-related providers and pioneers of innovative vehicle and mobility concepts. In addition to start-ups, it is suppliers and technology companies that give free rein to the spirit of invention and venture a new direction. There is the Waymo Car from Google [4] here, or the Apollo project from Baidu [5] there. In the search for innovation-savvy mobility consumers, the future clientele of buyers is maturing thanks to social media and direct marketing: be it events for test drives, the opportunity to gain virtual and physical insights into the progress of technological achievement, or the opportunity to become part of a new, purpose- and experience-bound community. Examples such as Riversimple [6] show how clever social media marketing, crowdfunding, and innovative technologies are combined—with the announcement of a new business model, curiosity is aroused even further and invites people to try things out and participate. Riversimple offers the use of hydrogen cars for a specific purpose in order to offer personal mobility as a service package in the future. To finance the production of ten vehicles, EU subsidies and capital were acquired through a crowdfunding campaign in 2017. It is interesting to note that anyone can participate as an investor with a stake of 10 British pounds [7].

All those who are preparing to win customers today for tomorrow's projects should also be aware of the impact of faulty and uncreative offers. An example: a German automotive company provides its customers with an app for vehicle monitoring with the purchase of an electric vehicle. However, the information services that can be called up on it correspond 1:1 to the information services of a fuel-powered vehicle. The app contains conventional filling stations, the consumption display only shows "litres" replaced with "kw/h" and innovative services are out of place.

There's a better way to innovate, isn't there? Isn't customer loyalty an essential component of the innovation process for those buyers who are already opting for an electric drive today despite higher planning and coordination costs and a short range? Aren't these buyers particularly suitable for feature testing and continuous improvement cycles? Shouldn't these buyers expect something more special than a simple copy of a standard application from another model range?

The innovators in the mobility segment have sufficient starting points for creating suitable test cases or gaining innovative features, be it the daily journey to work, the business trip to a sales manager's meeting or the monthly bulk purchase. Travel, event organisers, and shopping providers should provide their consumers with "mobility at all times" as an elementary part of their experience. Through a creative and clever mix of services and data, a clientele can be built up for the use of Mobility-as-a-Service in conjunction with a door-to-door experience. The data itself and data access required for this will be more comprehensible for the consumer due to the additional benefits presented, rather than simply recreating the check mark concerning the terms of use in front of the "Send" button in an online ordering process.

If, in addition, the consumer is offered the opportunity to receive a monetary or service-related fee for his data, the willingness of the donor and their commitment is increased. On the latter point, we refer to a genuine shareholder value that needs to be implemented. This aspect is highlighted in the third volume of this series. So much can be said about how our shareholder value approach is about a participation model: the consumer who makes his data available receives a revenue share from the profit that is gained by the data brokerage of the third party provider who makes money with our data.

Successfully Managing Innovations

The fundamental principle of successfully deployed innovations consists of the ability to act, legitimation and productive application. Being successful requires assertiveness and marketing skills—especially in organisational networks and in the internal competition of established companies.

Get Started and Join the Road

We from DoktorB helped to review dozens of projects, technologies, concepts, and strategies. In this publication we are introducing a number of third party add-ons: SAP® in Chap. 9, Alp.Lab in Chap. 10 and additionally the ENTOURAGE project in Chap. 11.

References

1. Flügge B (ed) (2016) Smart mobility – trends, concepts, best practices for intelligent mobility. Springer Vieweg, Wiesbaden
2. Moovel (2017) An urban mobility company, making cities smarter. https://www.moovel-group.com/en. Accessed 18 May 2018
3. Volkswagen (2017) MOIA social movement. https://www.volkswagenag.com/de/brands-and-models/moia.html#. Accessed 18 May 2018
4. Waymo (2017) Company website. https://waymo.com/. Accessed 18 May 2018
5. Baidu (2017) Apollo project website. http://apollo.auto/. Accessed 18 May 2018
6. Riversimple (2017) Company website riversimple – the future of sustainable mobility. https://www.riversimple.com/. Accessed 18 May 2018
7. Riversimple (2017) Company website riversimple – shaping the future of personal transport. http://riversimple.seedrs.com/. Accessed 18 May 2018

Smart Spaces

9

DoktorB Barbara Flügge and Christof Walter

Abstract

Using SAP® as an example, this chapter shows how sustainability can be lived and developed in one of the most central topics: electromobility. If interest groups and sustainability-driven political decision-makers are desperately looking for the right approaches to raising public awareness of sustainability and increasing the acceptance of electromobility, companies will implement them at long last. The vision of climate neutrality is taken into account by management requirements, company car regulations geared to employee needs, and the associated infrastructural measures. The decision by companies to present themselves as a mobility provider in future, in cooperation with customers and business partners, is of great significance!

9.1 Motivation

Barbara Flügge

D. B. Flügge (✉)
digital value creators (DVC), Bottighofen, Switzerland
e-mail: barbara.fluegge@dvcconsult.com

C. Walter
SAP SE, Walldorf, Germany
e-mail: christof.walter@sap.com

B. Flügge (ed.), *Smart Mobility in Practice: Good Car, Bad Car, No Car – Is this the End of Nurturing Our Mobility DNA?*,
https://doi.org/10.1007/978-3-658-43171-6_9

9.1.1 Creating Innovative Spaces

The innovation projects of aCar in Sect. 6.3 and Green Class in Sect. 7.2 are great examples of the use of electromobility and how elementary it is to have infrastructural preparation in place for the success of sustainable living spaces.

We are now focusing on the process of infrastructural equipment and commissioning called *Ecosystem Resourcing*.

> Ecosystem resourcing denotes the process of equipping the habitat with life essential, operation relevant, and functional resources. Thus, the process of turning the habitat into a self-contained functioning system is factored in. In the case of missing resources or backlogs in the supply and distribution chains resulting from contingency measures or natural disasters, the risk increases for the individual. Smart Mobility offerings function only with premeasures and operation concepts that are in place to combat connectivity outage in the field of telecommunications, energy provisioning, or climate related interferences. ([1], p. 196 f.).

This gives rise to an essential demand in the handling of cars: the demand for self-sufficiency!

Accordingly, the BIM, the building blocks of intelligent mobility, contain a large number of hardware-specific, technological, and communication-related elements under Smart Spaces that help to serve self-sufficiency. These are the first elements but there are additionally more elements to be added (Fig. 9.1).

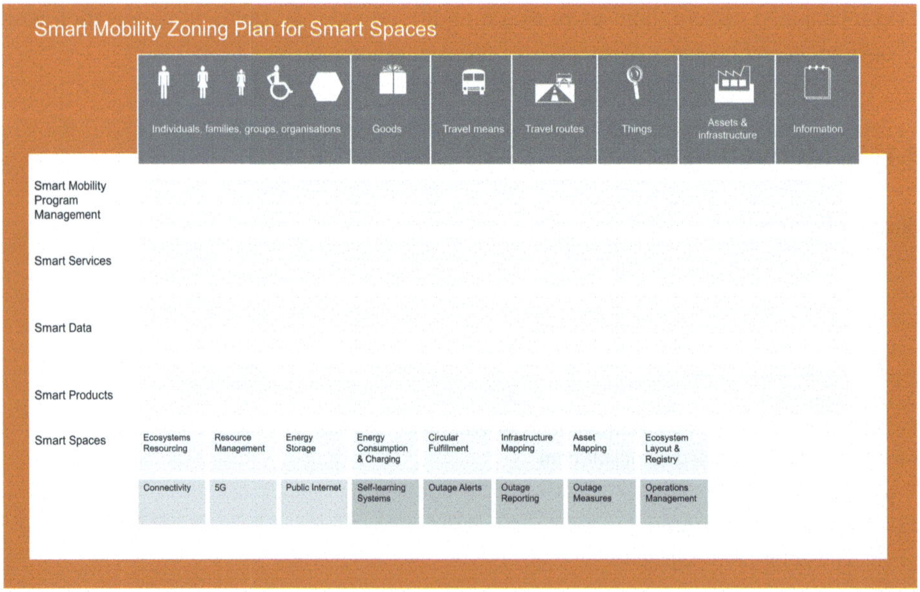

Fig. 9.1 Smart Spaces in the BIM. (From [1]; courtesy of Springer Fachmedien Wiesbaden GmbH. All Rights Reserved)

Today, the acceptance of electromobility in the passenger car sector lacks mileage reach, reliability in arriving at the preferred destination, tolerable charging times, easy handling and battery life, as well as the resolution of the social and environmental damage of the battery end-to-end life cycle. In addition, missing charging points and technical restrictions in connection with charging comfort and easily accessible charging points are further obstacles. The lack of focus on the supply of electric mobility in rural areas and incompatible charging devices do not exactly contribute to drivers' willingness to change. The financial funding directive for charging infrastructure for electric vehicles in Germany [2] opened up a funding volume of € 600 million until 2019, determined by the Federal Government from 2016, as well as established a second call with an application deadline of 30 October 2017. Those have been and still are too small to compensate the individual drivers' effort for detours (to get to the charging stations) and to plan time expenditure. Currently, we are informally aware of federal plans to cut the availability of electrical power in cities (for battery charging) driven by the shortage in energy supplies. The mishandling of finding new alternatives to combustion engines to serve our appetite for mobility urges a wake-up call for institutional and governmental repetition of institution to act. As demonstrated in the multi-year project with the European Union and Barbara Flügge' global "Mobility Moves Minds – build and grow again as a business" project, there needs to be the institutional desire (driven by their responsibility and empowerment) to execute a "holistic digital strategy driven by institutional and organizational leaders" ([1], p. 85 f.).

9.2 Creating Sustainable Mobility (Infra)Structures Using SAP®

Christof Walter

9.2.1 Corporate Strategies Under the Sign of Climate Change

Climate neutral by 2025! This is the goal of the software company SAP® [3]. Daniel Schmid, Chief Sustainability Officer at SAP®, describes the objective as follows:

> SAP®'s lever to shape a sustainable future is primarily how we enable our customers to achieve a positive economic, social, and ecological impact with the help of our products. But it has always been our ambition to lead by example and be a credible example of a sustainable company. [4]

Thirty percent of SAP®'s own emissions are generated by traffic-related business process shares. This includes those activities which range from the choice of the means of transport used by consultants, sales, and distribution, and management, the means of transport and tools used by the emergency personnel in the software production and maintenance process, and all such activities which can only be operated by means of energy expenditure. Examples of the latter category are canteen management, the operation of data centres, heat management, new construction, and conversion-related measures, such as the concluded construction of Building 49 in Walldorf, Germany.

SAP® relies on avoidance, reduction, and compensation to achieve its climate targets. The former is about avoiding energy wastage, for example through the use of virtual media and communication technologies that make travel, and thus the use of energy-heavy services, obsolete. Reduction results from the foresighted use of alternative, energy-saving media, intelligent building management and means of transport, as well as the joint use of energy-heavy resources. SAP® understands offsetting as the compensation of energy-intensive activities with the allocation of an internal, CO_2-based settlement model. External commitments are another cornerstone. This includes participation in the Livelihood Fund, which collects "funds for afforestation and climate protection projects in Africa, Asia and Latin America" [5].

9.2.2 Sustainability and Electromobility as Innovation Targets

In Germany, employees are offered a lease plan for e-bikes with a term of 3 years and a purchase option.

Sustainability plays an important role at SAP®. Employees in Germany can choose an annual train ticket called Bahncard 100. They are offered a company car from a wide range of vehicle models. Each year, the emission criteria (CO_2) and the limit values are pushed downwards and determine the vehicle models that are being offered. Particularly economical and low-emission vehicles receive three different environmental bonuses, and electric vehicles are also subsidised with a company subsidy. In the meantime, electromobility has become an integral part of the corporate strategy. As of February 2018, approximately 1250 electric vehicles are in use worldwide. An electrification share of 20% of the company car fleet is planned for 2020, i.e. 5000 electrically powered vehicles worldwide.

The useful life of the electric vehicle is 4 years. The share of electric vehicles in Germany is currently around 1000 vehicles, of which 80% are hybrid vehicles and 20% are full-current vehicles. In other regions, too, such as India, 43 electric vehicles are now being used with a range of 33 charging points.

In Germany alone, 250 SAP® charging stations are available. The charging points are pure SAP® charging points that are available to the SAP® workforce. The supply at home is currently carried by the employees. Pilot trials are currently underway with a limited number of users on the subject of home shopping. Solutions are considered that comply with tax and energy requirements without causing additional expenses on the part of the company. As far as charging points are concerned, identification management is already being considered in order to enable non-SAP users to use them, for example, after office hours and at weekends. The first implementation is planned for 2018.

SAP®'s goal of positioning itself as an electromobility provider is particularly well received at its headquarters in the Rhine-Neckar region. Employees realise the company's claim to implement climate neutrality in its operative business. Company car regulations and usage criteria create an incentive and reward conscientious use of resources. It is imperative to demand changed upper limits for CO_2 emissions or other mechanisms, and therefore employees are encouraged to rethink and act. The aim is to create a so-called green conscience. For business partners and companies or administrations, the continuous implementation of electrification on the part of SAP® is worth imitating. In addition, SAP® offers advantages for an entire region through joint voting rounds, co-innovation projects and employee ID management planning.

The chart at Fig. 9.2 shows how different organisations and companies at one location use networking via digital services and a coordinated charging infrastructure to provide citizens with an interesting range of services.

9.2.3 Electromobility as a Learning Field

Creating infrastructures and incentives relevant to electromobility through the public sector is at the top of many people's wish lists. However, organisations themselves can make a decisive contribution. Companies, vehicle suppliers, cities, municipalities, associations, and societies are now using electric mobility as a learning and experimentation area. In the case of utilities organisations, new service, usage and tariff offers, such as an intelligent energy mix for example, can be tested by employees and members of the workforce. Hardware and infrastructure providers use the opportunities to go through the innovation process themselves and with their customers, and to redesign it accordingly. The testing of new components, various test series with components and innovations, and changes in the organisational operational sequence thus become a component of the project. Similarly, companies in the IT and software industries will learn how to handle hardware and the variety of activities involved in setting up, operating and maintaining charging points, billing and structural planning for space requirements, and approval processes.

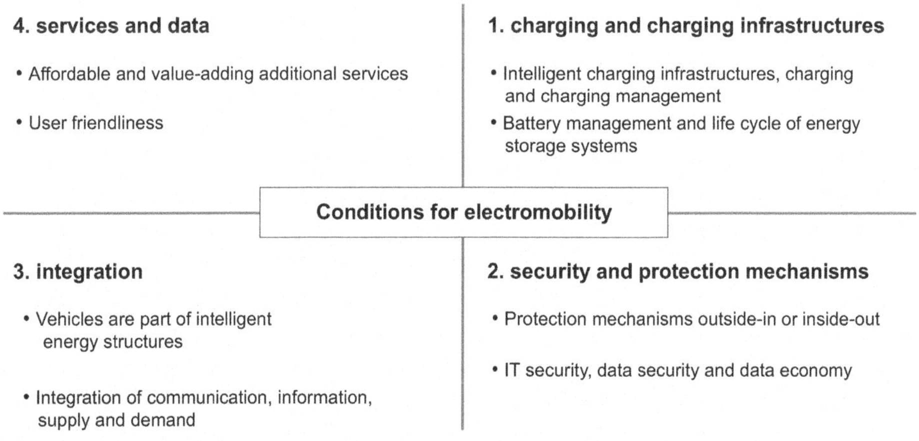

Fig. 9.2 How mobility service providers could promote electromobility in their own companies

When researching new functionalities or markets, SAP® consciously relies on joint projects with stakeholders and forward thinkers. Thus, representatives of cities and companies, and science and research institutions can be found in the co-innovation projects. Closing ranks with complementary partners such as the energy supplier EWE AG, hardware and infrastructure providers, as well as decision-makers from the public sector, is based on the goal of creating sustainable mobility structures. The Mayor, Chief Digital Officer (CDO), Executive Board, Chief Innovation Officer (CIO), Sustainability-/Fleet Officer or fleet manager and scientists agree!

9.3 Opening up for New Markets

Barbara Flügge and Christof Walter

9.3.1 The Vehicle as a Sustainable Space

The lack of acceptance of electric vehicles is not due to a lack of willingness on the part of buyers or vehicle renters to protect the climate and act sustainably. The financial incentive for the purchase of an electric vehicle in Germany, for example, is not decisive for the low sales figures. Lack of range, doubts about the longevity and robustness of vehicles and batteries, the comparatively high costs with manageable performance, the insufficient number of charging stations and the high planning and time expenditure make acceptance difficult.

Companies such as SAP® are taking on a pioneering role in this innovation-driven environment. Due to the high number of vehicles in the company car pool, the company also becomes a first customer who can submit suggestions to vehicle manufacturers, infrastructure planners and others. These are available as sparring partners. Especially in business associations and regional interest groups, it is worth being prepared to share one's own experiences and challenges in dealing with electromobility with others. If several companies were to form alliances, far more citizens and employees would be motivated to take part in electromobility: metropolitan regions and regional alliances would be able to generate added value for their populations by means of infrastructure support, car sharing and a sufficient number of charging points and digital additional services. In addition, own mobility offers for business travellers can be offered. Especially with this clientele, the comfort experienced and the benefits generated from it quickly spread!

Similar to SAP®, sustainability programmes such as the Green Class from SBB, BMW, and PubliBike are based on the functioning electric mobility infrastructure and the growing acceptance of drivers and passengers. The expectations of the experience of being on the road are met with the services offered. The service package is one of the key factors from which SAP® employees do benefit. This includes the free choice of car make, fuel cards, the collection of vehicles on site and the elimination of unnecessary walking and driving paths. The range of services, which includes tyre changing, tyre storage and workshop services, is comprehensive and compensates for the in-house effort in the electromobility segment to take care of the charging infrastructure at home.

In the future, digital meters or mobile charging processes can be considered. An example of the latter is Ubitricity [6]. This also allows further infrastructural measures by the public sector, such as intelligent street lighting, to be reconciled. Set-up, maintenance, and operational expenses are to be considered separately within the framework of a cost/benefit analysis, as are expenses with regard to the space required for the loading process at the roadside.

> **Electromobility Creates Space** The transformation of the vehicle space into a living space stands and falls with a functioning infrastructure and service offer, which is characterised by alternative, sustainable engines.

9.3.2 Structured Acceptance Management

In 2018, SAP® kicked off a plan for a broad range of services with its partners. Companies and administrations benefit from a one-stop offer ranging from fleet conversion, installation and operational specifications to the optimisation of the e-fleet solution. The integration of storage and photovoltaic solutions takes place as required. SAP® would not be a solution provider if the development of a standard software solution for an open system for all manufacturers, all models of vehicles and charging infrastructures for mixed fleets were not also part of the catalogue of ideas. In addition, experience from load distribution, for example in rail transport, could also be transferred to intelligent charging and load management for electric vehicle fleets.

Demand-driven loading and the distribution of available services to other vehicles play a major role in the acceptance of fleet management. Further user groups are opening up with growing offerings. In addition to our own fleet management and staff, this includes emergency personnel and employees of business partners and customers. Decision-makers from companies, investors, leasing companies and OEMs who take part in the SAP® Roundtables on Electric Mobility use the opportunity to transfer knowledge, share project planning experiences and join forces. The latter is not only about monetary support, but also about the provision of skills and experts. Here, the planning of a demand-driven service offering also benefits those who are at certain geographical distribution and loading centres, and who will benefit from overcapacities and other services through the use of the planned software solution. By means of real-time analyses, the vehicle itself and its users optimise use, charge, and energy management as well as reduce costs and CO_2 emissions safely.

The cross-company interest in bringing flexibly designed, yet coordinated, infrastructure offerings to the market will make itself felt amongst end-consumers—whether employees of one's own or another company. A foreseeable result is nationwide loading points and a one-stop shop offer tailored to third-party suppliers and consumers. This also requires swift implementation, further comrades-in-arms and public support. Sponsorships and the connection of ridesharing solutions such as TwoGo from SAP serve this expansion within the geography. To this end, further branches and locations of a company are to be successively included in the infrastructure and service planning.

9.3.3 Scalability by Strengthening Collaboration

The above-mentioned steps with regard to roundtables, the use of local and supra-regional contacts to business partners, and the deliberate involvement of customers, help newcomers such as SAP® to establish a good basis for the expansion of the existing range of electromobility. SAP® is a newcomer because the company has not yet been a mobility provider in the sense of Smart Mobility. The odds are good. The staff and the ecosystem of partners, consultants and service providers are innovation-savvy and curious to try out new technologies themselves and, with the appropriate expertise, to improve them. The geographical placement of locations and branches and a workforce that, on the one hand, has a high level of travel activity and, on the other hand, carries out a not inconsiderable proportion of the work from home, allows sufficient critical success factors and variants of intermodal offers to be derived! The travel activities and the city/country distribution of the workforce make it possible to carry out a variety of infrastructure tests. SAP® conducts appropriate employee surveys here.

Further target customer groups will be derived from users of offers similar to the Green Class of the Swiss railway organisation. The focus here is on those who are in the decision phase (1) for or against a car and (2) for or against an electric vehicle. Today, electric mobility is regarded as one of the core technologies for the use of autonomous vehicles. In addition to the following customer groups, there are other customer groups here, such as frequent flyers. This is because frequent flyers are used to relying on automatisms and autopilots in air traffic. We apply this analogy in Sect. 12.3 as an argument for an ecosystem embracing mobility management.

I ask Christof about his point of view on hybrids. "The hybrid approach for company cars", says Christof Walter, "is not intended: often, when the range is short, there is no time to look for a charging point, or simply out of convenience, the additional charging is activated via the combustion engine. This is not the intention of the inventor, especially since the additional charging is intended for an emergency and not the rule. The CO_2 emissions are immense". Given these insights, shouldn't we educate about the uselessness of these hybrid models and help to concentrate on better infrastructures and end-to-end life cycle operations? "Exactly! The infrastructural and simplified planning of when, how and where the charging stations are booked and used is an important prerequisite for a higher level of electromobility", Christof Walter explains. He continues: "The other prerequisite is the conscious use of our resources". The change of consciousness is not only necessary from the employee's point of view, but also from the company's point of view. It is for nothing that the pool vehicle, which can be called up immediately when fully charged, likes to serve as a metaphor for intelligent mobility? As a fleet manager of a company or public administration, very quick decisions are required: (1) Which means of transport are used? (2) Is Carsharing or Ridesharing with TwoGo available? (3) How can resource bottlenecks be solved in a creative and environmentally friendly way?

And how does Christof see the addition of other offers outside the current votes? For Christof Walter, "it needs the application of a standard in order to be able to use and offer mixed fleets in the future. Otherwise, the infrastructural expenditure for companies and households could not be calculated. The goal of SAP® is to implement the European Union Standard, named OCPP 1.6". OCPP stands for Open Charge Point Protocol.

OCPP 1.6 transforms the charging station into an intelligent system that allows distinct software functions to be bundled and exchanged. The integration of equipment from distinct manufacturers is possible. A higher usage rate is expected through the openness of the technology. The OCPP 1.6 standard (Open Charge Point Protocol) was updated by security guidelines (see OCPP 2.0.1).

Christof moves on, "in electric mobility and a networked electricity market in particular, standards are fundamentally the basis for viable solutions that we want to support with our software".

Standardisation leads to increased acceptance and the bundling of investments. Instead of isolated and incompatible systems, budgets and project plans that commit themselves to a standard come up against the expansion of production areas and event locations in the urban and rural ecosystem. Those who were not convinced to date and were afraid of the monetary and planning effort will find it easier to access the infrastructure. With higher performance, brand-conscious and sporty drivers will also be among the users in the future. According to Christof Walter, the brand selection will be made at the national level. Among the pool vehicles, i.e. those vehicles that can be booked by employees for short-term assignments, electric vehicles are also available. For example, this might be a Tesla vehicle in Switzerland or France. Triggered by a business trip, drivers and passengers are given the opportunity to experience e-mobility and make up their own mind.

Passengers who have already decided on ridesharing within the company framework of sustainable mobility, will welcome additional offers, provided they offer them the appropriate experiences in the vehicle.

In the spirit of the Smart Mobility ecosystem, we foresee further interest groups that will be joining:

- Content providers for the processing and distribution of content such as the reach of specific shopping routes in rural and urban areas
- Aggregators that bundle services according to a specific purpose or experience
- Sales of service packages, e.g. for event visitors
- Cloud providers for providing, processing, and securing large amounts of data in and around the car

- In the future, we from DoktorB project a completely new experience consulting business that will determine how the consumer markets responds to smart mobility needs. Tailored to the needs of new living, learning, and work spaces, we see an upscale and reintegration of regional mobility providers. Services and goods providers will be able to experience a new sales market that is truly mobile!

9.3.4 Outlook and Estimates

In the coming years, electric mobility will develop more progressively and more comprehensively than many predicted a few years ago. The use of new, innovative battery technologies is still a long time coming, but an intensification of research and development investments can be observed worldwide. According to current figures, the stock increased by 55% compared to the previous year [7].

The information, data and telecommunications networking of vehicles, infrastructures and buildings is a topic that will be massively expanded in the near future. By using photovoltaic systems with intelligent battery storage in private and business buildings, energy can be generated and stored for other periods. As a mobile battery storage device, the vehicle can not only absorb energy together with a stationary storage device but can also return it to the energy supply system with the appropriate technology. A large fleet of companies is thus able to cover the energy requirements of the networks much faster and more cost-effectively than a hydroelectric power plant connected to the grid several hundred kilometres away.

Due to the much lower complexity of an electric motor compared to a combustion engine, an electric vehicle requires less maintenance, is less susceptible to damage and has a longer service life. In addition to dispensing with a highly complex multi-speed transmission (electric car: 1 gear), an electric motor has an average of approx. 250 parts, including screws, nuts, etc., compared to 2500 parts in a new generation combustion engine [8]. The very low volume at high acceleration values additionally distinguishes the electric motor. However, with regard to efficiency, one still has to differentiate at present. With an efficiency of approx. 90%, the electric motor is comparable with the efficiency of a high-performance internal combustion engine, taking into account the currently prevailing energy chain mix of generated and stored electricity. If, on the other hand, the electricity is generated in an environmentally friendly way, the efficiency increases [9]. Conversely, this would also mean that modern mobility requires environmentally friendly and sustainable energy generation and storage.

Particularly in areas of the world with poor infrastructure or poorer infrastructure, the location advantage of environmentally friendly energy generation is often possible. Morocco generates 160 megawatts of energy from a gigantic solar plant with 537,000 mirrors—making the desert usable. An expansion up to 560 megawatts is currently being implemented. The situation is similar with wind and waterpower. The aim of an intelligent and sustainable mobility policy should be to make existing energy sources such as the sun, wind and water usable and storable. This allows a transfer to more mobility intensive areas. This results in new business systems that make the energy and mobility market massively more flexible!

9.4 DoktorB's Checklist

Important Basic Considerations

- Which companies and stakeholders with a similar approach are you interested in?
- What does your company look like? What does your own life cycle assessment look like?
- How often do you/your employees drive a car? And how many kilometres?
- What's the longest stretch you've driven this year?
- How much time per month do you spend refuelling, maintaining, and servicing your vehicle?
- How much time do you spend on reservation, booking, and payment?
- Do you use a mobility application, an app outside the vehicle's navigation system?
- What improvements do you want for the app?
- Does the functionality of your vehicle (navigation, information services, use of apps outside the vehicle) influence your decision as to whether an electric vehicle is an option or not?
- How many charging stations are available on the way to work or within a radius of 80 kilometres?
- What information do you need in advance before you would travel with an electric vehicle? Be very thorough one day on the way to work or to the nearest town:
  1. Take a look at the vehicle cockpit.
  2. What offers do you notice along the way for loading, maintenance, and repair?
  3. What information should be suggested to you before you leave the house?
  4. What information should be suggested to you during the journey?
  5. What feedback or recommendations do you expect when you park your car at your company or the place of arrival?
- What personal information are you willing to provide to an app that offers round-the-clock support on the five points mentioned above?
- What assistance does your current vehicle offer you: a hotline, a message on the display, a message to the app?
- Your company is planning a 50/50 share of electric vehicles for the vehicle fleet: what input do you give the project team?
- How high are your fleet costs at the moment and how are the costs divided into systems, operation, maintenance, acquisition/leasing, vehicle tax and insurance, repair and replacement vehicles?
- How do your expenses change when upgrading electric vehicles? Are you planning to use Mobility-as-a-Service?
- What advantages does a procedure like SAP® offer you with business partners and customers on a regional level?
- What considerations have you made to sensitise employees to the sustainability goals? What services do you plan to offer your employees in the vehicle to make it easier for them to switch to an electric vehicle—even if only for the journey to the next sales appointment?

References

1. Flügge B (Hrsg) (2017) Smart mobility – trends, concepts, best practices for intelligent mobility. Springer Vieweg, Wiesbaden
2. The Federal Government (2016, May 18) Improved promotion of electric vehicles. https://www.bundesregierung.de/Content/DE/Infodienst/2016/05/2016-05-18-elektromobilitaet1/2016-05-18-elektromobilitaet.html. Accessed 18 May 2018
3. Riversimple (2017) Company website riversimple – shaping the future of personal transport. http://riversimple.seedrs.com/. Accessed 18 May 2018
4. SAP (2017) New target: climate-neutral by 2025. https://news.sap.com/germany/klimaneutral-bis-2025/. Accessed 18 May 2018
5. Livelihood Fund (2017) Company website. http://www.livelihoods.eu/. Accessed 18 May 2018
6. Ubitricity (2018) Company website. https://www.ubitricity.com/. Accessed 18 May 2018
7. Ecomomento.de (2018) The number of electric cars and plug-in hybrids is rising significantly worldwide. https://ecomento.de/2018/02/15/zahl-der-elektroautos-und-plug-in-hybride-steigt-weltweit-deutlich/. Accessed 18 May 2018
8. The Electric Car Journal (2017) Electric motor vs. internal combustion engine. https://e-auto-journal.de/elektromotor-vs-verbrennungsmotor/. Accessed 18 May 2018
9. Auto, Motor und Sport (2018). https://www.auto-motor-und-sport.de/news/effizienz-wie-effizient-sind-elektromotoren/. Accessed 18 May 2018

Innovative and Autonomous: Test Passed! 10

DoktorB Barbara Flügge and Horst Bishop

Abstract

Innovation through collaboration! Since its foundation, the Automotive Cluster Styria, AC Styra for short, has proven how collaboration and innovation have a lasting positive influence on the competitiveness of an entire region. The Alp.Lab project now faces the task of establishing itself as a test laboratory for autonomous driving. The public authorities gave the go-ahead for test laboratories. Supported by the Ministry of Transport, the aim is to test innovations in-house and to assert oneself as a collaboration partner for newcomers and disruptors on the world market. Whoever takes which position, this process demands openness and a lived discourse.

10.1 The Alp.Lab Test Laboratory

Innovative mobility offers, which rely on automated driving as an admixture of a modern mobility mix, hope for an advantage in terms of saved space, reduced infrastructure demands and maintenance, less traffic disruption and congestion, as well as a departure from vehicle ownership. This is the forecast for autonomous driving in urban living space. Others speak of an increase in traffic density due to the elimination of disused or

D. B. Flügge (✉)
digital value creators (DVC), Bottighofen, Switzerland
e-mail: barbara.fluegge@dvcconsult.com

H. Bishop
Graz University of Technology, Graz, Austria
e-mail: bischof@icg.tugraz.at

B. Flügge (ed.), *Smart Mobility in Practice: Good Car, Bad Car, No Car – Is this the End of Nurturing Our Mobility DNA?*,
https://doi.org/10.1007/978-3-658-43171-6_10

depreciated city buses and the use of autonomous vehicles compensating for these buses. Autonomous vehicles can react more quickly to short-term requests and can also be used in small and off-route streets. However, in order to gain insight into the predicted benefits, assumptions, and simulations are insufficient. The planning units of the public sector have the task of investing for society in long-term sustainable infrastructure measures and mobility projects in order to make such effects comprehensible and assessable. Decision-makers would then have access to reliable insights and concept proposals. How much advantage results from automation or autonomous driving then depends on the targeted location and its location profile.

Extensive tests and trials are needed to bring technological reliability and social acceptance into the focus of public discourse. Technological reliability is based on the smooth and secure interaction of new and proven technologies in all eventualities. The illustration Fig. 10.1 compiles the technologies that are essential from today's point of view and which are absolutely necessary in connection with successful autonomous driving.

The alignment of today's digital vehicle systems undoubtedly requires disconnecting from the original purpose of the steering wheel. The classic alignment of front and rear seating areas is also dissolved. We will encounter increasingly cell-like development of vehicle shapes. Passengers and co-drivers of autonomous vehicles experience the same co-travelling feeling as bus and train passengers. We have conceptualised an ambient living assistant in order to maintain the interior, which should remain adaptable to the wishes and habits of the passengers. The digital assistant will be used by consumers throughout the day, with or without a car even. Upon a guest's booking, the autonomous vehicle processes the guest's preferences and profile, which is uploaded to the digital concierge, and pulls in the content for entertainment, education, and recreation (see Sect. 4.2).

The equipment of the interior space and its functional orientation go hand in hand with the outdoor parameters. These include road and path structures, as well as building

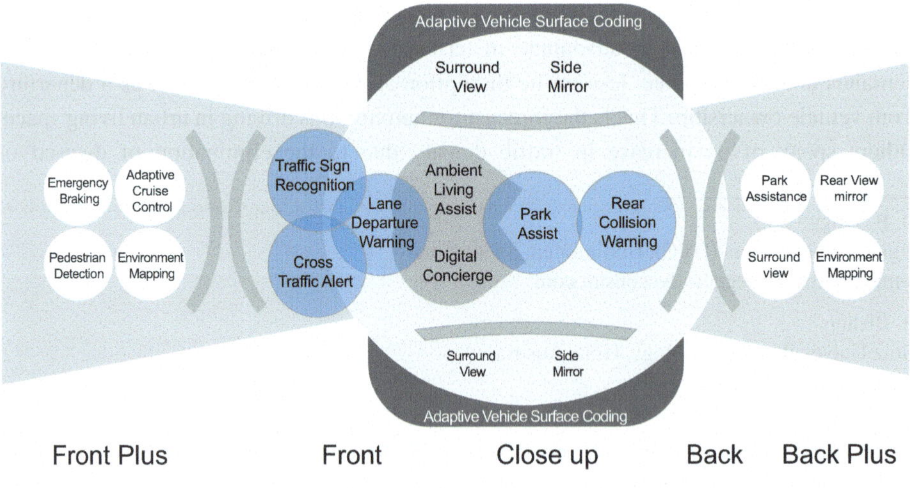

Fig. 10.1 Technology in use for autonomous driving

decodings and the preparation of digital signal points on footpaths, crossings and crossings for access to workshop, maintenance and car park locations.

In order to fulfil the social mandate, co-innovation projects, local and regional initiatives, and test regions need to offer a broader spectrum of citizen participation and living labs.

The history of Alp.Lab [1] in Austria is a successful example of a planning approach supported by the public sector. Alp.Lab stands for Austrian Light Vehicle Proving Region for Automated Driving. A study by AustriaTech, commissioned by the Ministry of Transport BMVIT, dealt in 2016 with the mobility of the future and the focus on autonomous driving [2]. In addition to further proposals, a call for tenders was initiated for the launch of an independent company for test series and the establishment of test regions for autonomous driving. The region of Styria, home of the Automotive Cluster (AC) [3], is predestined for this. More than 290 companies from the automotive, aviation, and railway industries belong to the AC [4]. In addition to around 55,000 employees in the above-mentioned companies, the project addresses public institutions and municipal administrations located in Styria, particularly the city of Graz, research institutions and innovation centres, as well as the population living and working in Styria.

At the time of the application process, the Technical University of Graz [5] was asked to manage the tender and therefore to act neutrally. Essential framework conditions of the tender stipulated the establishment of a company with limited liability, a conclusive business plan for the transfer of the research activities into a self-supporting business entity, the co-financing of a test region and the necessary infrastructure, and a tailored service profile. In addition to formal criteria such as the foundation of the Alp.Lab entity, competence needed to be made available in the field of simulation and data acquisition, taking into account the use of autonomous driving on public roads. In addition, existing or representable know-how with regard to test and simulation procedures was essential.

In 2017, following a successful call for tenders, Graz University of Technology, AVL List, Magna Steyr, the VIRTUAL VEHICLE competence centre, as well as Joanneum Research, founded Alp.Lab GmbH [6]. Alp.Lab received a total fund of 5.6 million euros from the Austrian Ministry of Infrastructure, which covered the ALP.Lab test environment and two research projects (ibid.). Supported by an interdisciplinary network of about 50 organisations, Alp.Lab GmbH is supported by a consortium of experienced members with an affinity for technology and innovation. Major partners are, for example, ASFiNAG [7], which, as Austria's motorway operator, provides the necessary public test infrastructure such as roads and tunnels. The city of Graz is an important partner, serving as a lighthouse stakeholder.

10.2 Digital and Physical Tests

The heart of automated driving lies in the intelligent and successful interaction of innovative and proven technologies. A continuous digital test chain is an essential unique selling point, which Alp.Lab is committed to. Eliminating breaking points and pointing out improvement potentials sounds simple.

It becomes complicated when it comes to researching the causes and identifying the technology(s) involved and their representatives: (a) at which point in the test chain something was changed or updated, (b) due to which technical or structural constellations research results changed or did not come true, and (c) under which conditions changes in driving and control behaviour occur. The latter applies not only to the vehicle, but also to components, test and inspection procedures and data sources from a wide range of sources—within and beyond the car.

The execution and evaluation of tests does not only require technical and digital know-how. It also requires organisational and clarification skills, a cross-organisational business model geared to Alp.Lab and a functioning back office for purchasing, accounting, and controlling.

Alp.Lab initially focuses on automated and autonomous driving for individual traffic, i.e. cars, on public roads. Primarily urban and rural areas are used here. Alp.Lab customers have public test tracks as well as private test tracks at their disposal. Availability and usability of essential technologies and test infrastructures are an essential factor for the success of Alp.Lab.

Let's take a look at Alp.Lab's service portfolio in detail [8]:

- Simulation environments
- Driving simulations
- System tests of driver assistance systems
- Sensor Validations and Qualifications
- Tests on public and private roads
- Manoeuvre tests
- Data-specific links and cloud-based services
- Logical "acceptance tests" of components

Alp.Lab relies on comprehensive support for this. This ranges from scenario-specific consulting to test preparation, execution, follow-up, and evaluation. Alp.Lab cooperates with other test laboratories in Europe, such as the A8 motorway and test tracks in Düsseldorf and Dresden, both in Germany, the A36 in France, and further test regions in Germany and France. In Germany, these include Kempten, the Free and Hanseatic City of Hamburg, the federal states of Baden-Württemberg and Lower Saxony, and in France, for example, Strasbourg.

The unique selling point of Alp.Lab in comparison to other test laboratories lies in the mapping and execution of complex and complete test series and a comprehensive range of services. The composition of the Alp.Lab consortium plays a particularly important role here.

Let's look at Alp.Lab's offer from the point of view of a sensor manufacturer who wants to use their services. The aim is to investigate the suitability of sensors which support functional areas as described in Fig. 10.1. Integration and testing tests are expected. Test tracks are available for the latter. These are predefined by Alp.Lab. In addition to simulation tracks and simulation tests, tests are available on public roads through cooperation with ASFiNAG and on private test tracks by consortium members. Depending on the test series, vehicles are available for the latter. In clarification with the operators Magna Steyr and AVL an operation takes place. Various tests such as individual routes or round trips are conceivable.

What is actually interesting for test laboratories is the specialisation in subject or functional areas and test types. In the case of Alp.Lab, the advantage is that, in addition to the regular traffic of autonomous vehicles, exceptional situations can be taken into account. Specific scenarios are elementary, such as tunnel trips, trips in mountain regions and trips with changing seasons and weather fluctuations. This includes winter tests and temperature tests, as well as the inclusion of access roads for toll roads, company, event, or construction sites that are relevant for approval.

Further tests are required in low temperature and icy regions, and regions or areas with limited visibility or a high throughput of vehicles at transshipment points. Not to be neglected in the future will be the mapping of business processes, for example through changing the rental or ownership conditions of an autonomously used vehicle, as well as billing and data protection mechanisms, to name two examples in mobility operations. A further benefit comes with the evaluation of the measured data and providing data to component suppliers, route planners and research laboratories! It will be interesting to see to what extent test data transforms into a tradable commodity and how results impact the profitability and establishment of Alp.Lab as an independent company.

Alp.Lab does not currently carry out tests in city interiors. Testing in the city of Graz is planned [8]. Scenario selection and the definition of test environments, test parameters and target definitions are carried out by sensor manufacturers and Alp.Lab together. Alp.Lab takes on the role of the service provider. The clarification of legal procedures is just as much a part of the service portfolio as obtaining test approval.

As a full-service provider, Alp.Lab is available to the sensor manufacturer as a central point of contact and intermediary. Due to Alp.Lab's organisational and membership structure, the company remains manageably lean. Marketing measures were not necessary in the beginning, but over the past 2 years we observe a greater effort to be heard—that is one outcome of the competition in the Asia-Pacific and Pan-European geography. Support is managed following the interdisciplinary ownership structure of Alp.Lab and the network of the consortium spanning research, industry, business, and public administration. On the other hand, in relation to the membership at AC Styria, the network approach has shaped the network-oriented thinking and actions of private and public cluster members. Conversely, companies, customers, and business partners of the AC Styria Hub know the Alp.Lab. Owners, such as the University of Technology Graz, profit from an increasing research demand to build up competence and thus create a further unique selling point in international competition. The Graz University of Technology, for example, conducts research and surveys on the autonomous delivery of mail in the city of Graz.

10.3 Growth Through Early Position Determination

10.3.1 Networking Right from the Start

The aim of the investors is to continue Alp.Lab as an independent and self-supporting company with limited liability, a GmbH as it is called in Austria, and to establish itself as an incubator and service facility for automated and autonomous driving in the local and international market. This gives Alp.Lab the chance to become a de facto standard for testing autonomous vehicles.

Through its consortium and partner structure, Alp.Lab has created a complex but also suitable basis for positioning Alp.Lab Services in the market and gaining new customers through its networking. It would be more than interesting to look behind the scenes to see the logical and organisational orchestration mechanisms. In particular, this is useful when it is necessary to document logical and technical load tests for audit purposes consistently across several components and thus achieve traceability of processes.

In the medium term, companies such as Alp.Lab will provide legislators and political decision-makers with the opportunity to derive basic principles and findings for regulations and cooperation in the field of autonomous driving. The participation of TÜV Austria [6] as a partner of Alp.Lab has created the conditions for preparing certification services and making them available to the national and international market in the future.

10.3.2 Data, the Tradable Good

A further gain in knowledge for all participants is the handling of data. Test results or data obtained as tradable goods are not only important for the acquisition of new customers or as additional business for Alp.Lab. Two questions arise at an early stage in the chain of test series within the framework of the contract to be offered: (1) Who owns the data and (2) Which data must be viewed or transmitted for testing? If, conversely, assumptions can be made without data transmission, the question arises as to their resilience in normal and exceptional cases. If data from a customer or a component is required for test series, this could result in an exchange of data for test execution. In the third volume of this series, we ask ourselves questions about profitability and growth potential.

10.3.3 Test Spaces

It will also be interesting to observe how public administrations in cities and municipalities position themselves in relation to autonomous driving in general and Alp.Lab in particular. For municipalities and cities, the focus is not only on inner-city tests, but also on making road sections and access roads available. This results in real learning and testing opportunities for citizens and decision-makers. The resulting experience advantages cannot be compared with those of test laboratories that focus on simulations and tests with model construction vehicles.

Alp.Lab's ability to grow is to be found in the balance between domestic on-site projects and the testing of realistic end-to-end simulations in a laboratory environment. In our opinion, this requires a laboratory environment that is digitally accessible to foreign interested parties. This would be the first milestone for a qualitative assessment of the services, from definition, execution and processing to evaluation and billing. Foreign interested parties thus have the opportunity to create the basis for on-site tests on the domestic market. Climate technology, topographically relevant legal bases and manners play an important role in the participation of administrations, citizens, and companies. Like an offshore provider of programming services in the IT industry, there is a need for clear digital boundaries per test customer and per test object; in addition to data and processing processes, authentication and data transmission mechanisms are required to secure the respective data and test room and the access to it, and to establish trust for the use of data.

10.3.4 Test Autonomy

The autonomy of tests is limited when it comes to testing in physical, namely public space. Assistance is provided by agreements on predefined roads and intersections [9] as well as autonomous driving with an accompanying person [10, 11]. Private test tracks have become more popular. Alp.Lab co-owner Magna Steyr, for example, also offers private test tracks.

In South Korea, a similar test site, the K-City [12], is currently further under construction following its predecessor, the M-City in Michigan, USA [13]. In South Korea, K-City spans an area of 300,000 m^2 which enables autonomous driving tests in urban space.

The question already arises as to when autonomous driving can be booked as a holiday experience in K-City. An autonomy of tests is already taking place today through functional recordings behind the scenes—given the consent of the vehicle owner, the person who books the vehicle equipment or who approves and pays for the upgrade. Similar to behaviour data when shopping and installing apps on the smartphone, user behaviour and usage can be checked and admitted. Users can receive further offers. Monitoring driving behaviour, acceleration and braking, resource consumption, the use of certain entertainment functions and popular routes, and stopping points all help to achieve this. If driver profiles can be compared with social profiles and drivers' shopping profiles, the matchmaking of demand and desire serve as indictors and help us to craft new offers.

Extensive tests are necessary in order to learn further and to guarantee a high security of the module to be offered or the planned functionality. In addition, further proven components will be added and will replace even previously proven components. A reliably tested continuous chain may then show a different behaviour under newly created conditions. This leads not only to the questioning of the knowledge gained, but also of the underlying test and evaluation procedures. Cooperation between the partners is also required. Depending on the degree of maturity of a component, this has a promoting or inhibiting effect on the overall complex. The example of Tesla and Mobileye illustrates the fragility of a partnership, both internally and externally:

> "The latest critique of Tesla's Autopilot platform comes from the analysts at Navigant. The research firm just released a report on the automated vehicle market, and it ranks Tesla last out of 19 companies. Tesla scores high for its vision, but they haven't demonstrated the ability to consistently execute that vision", explained Sam Abuelsamid, senior analyst with Navigant Research and co-author of the report. […]
>
> A major part of that regression comes from Tesla's public split with its early partner Mobileye, an Israeli chip and software developer focused on computer vision tech. Mobileye was acquired by Intel for $15.3 billion well after the split with Tesla. Tesla worked with Mobileye on the original autopilot. But in the spring of 2016, a driver died in a Tesla car following an accident during which Autopilot was engaged. Tesla's autopilot was investigated, but later determined not to be at fault. [14]

To remain in the market as a test laboratory and to be able to justify their investments to investors and supervisory bodies requires the secure handling of a multi-layered network of providers, political decision-makers, stakeholders, and influencers. Networking work is needed both digitally and physically, on the domestic and international stage.

10.3.5 From Platforms and BIM

We encounter test autonomy when it comes to system tests and integration tests. The availability of a test laboratory plays an important role in this. The core is the presence and access to the individual components, which is a critical success factor, especially due to the large number of business partners, decision-makers, and contributors. The foundation of this is the Modular Integration Platform. The individual physical and IT-technical components of the consortium members are logically combined and linked. Alp.Lab takes over the orchestration of the integration platform and its operation. It is worth making a comparison with the reference architecture for Smart Mobility, which we set up as part of [15]. This can be found in Sect. 4.5. in Fig. 4.1.

Let's take a look at the BIM (Components of Intelligent Mobility): the following *BIM* are used at Alp.Lab, Figs. 10.2 and 10.3 as outlined with the black lined colour boxes.

Let's remember that *BIM* are functional and technical elements that result in the development plan for mobility projects. We make BIM available for project and change management (Smart Mobility Program Management), services (Smart Services), data (Smart Data), products and solutions (Smart Products), as well as physical and virtual spaces (Smart Spaces). For the latter, see also Chap. 9.

Another aspect which is covered in the *Smart Mobility Building Blocks* is the Digital Concierge. Once a city promotes the use of the *Digital Concierge* for all citizens, characteristics such as *signal points, movement related parameters, traffic behaviour, and safety measures* will be compared with the means and operations of transport and its corresponding data. Data analytics and recommendations kick in as soon as citizens are moving towards publicly and digitally retrievable places, event locations, and other kind of mobility gatherings. A similar service is conceivable for visitors to a city. This could also be well applied to moving assets, goods and installations and mobile, autonomously moving cells for goods, delivery robots, drones, and other forms of movable items (see Sect. 12.3). In the movable space and the space where we traverse, and connect, a relationship is built between modules/vehicles and the environment. It is the environment that provisions information, critical data and influencing factors and conditions, so that the control of the autonomous vehicle is thus enriched and further improved.

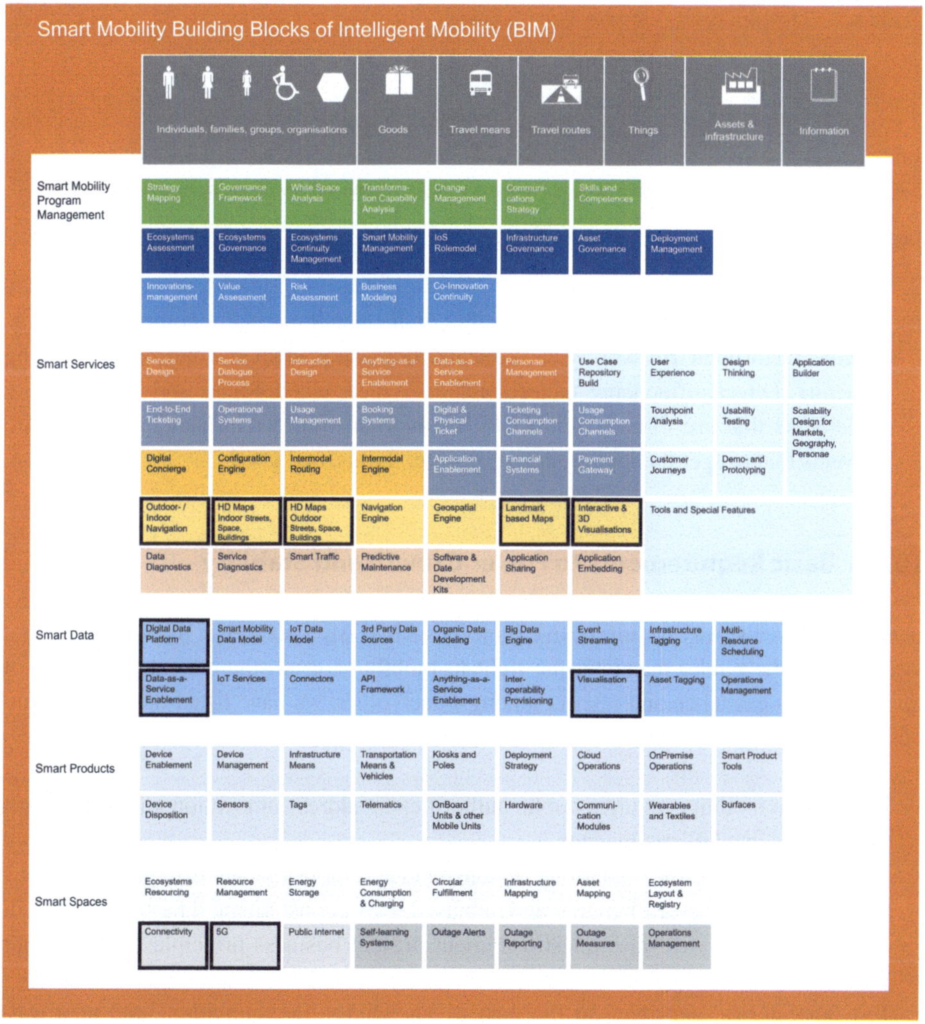

Fig. 10.2 BIM devices in use (1)

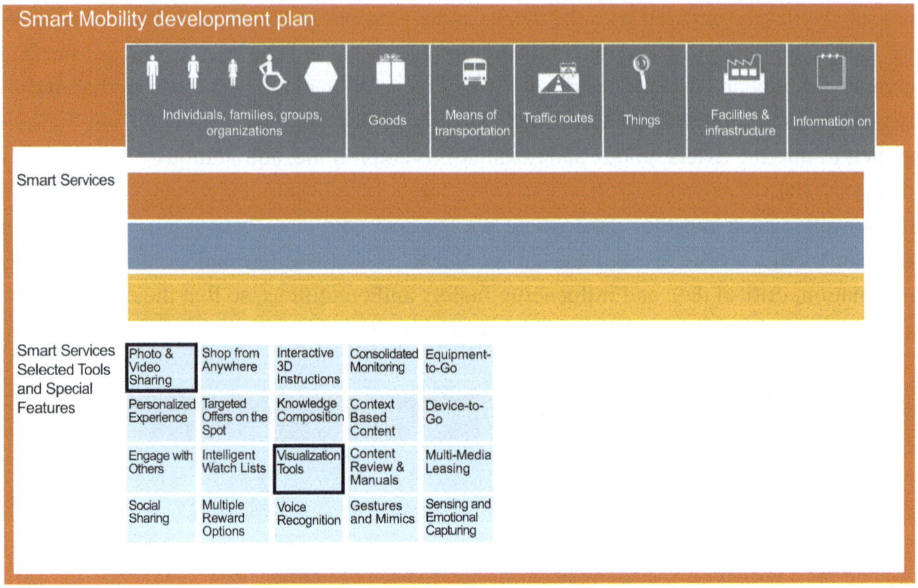

Fig. 10.3 BIM devices in use (2)

> **Safe and Autonomous on the Road** Safety in the handling of autonomous vehicles is achieved through trustworthy and ethical action using direct and indirect influencing factors. Direct influencing factors include vehicle technologies and infrastructure elements that are directly related to locomotion. Indirect influencing factors are events that trigger the locomotion of people, things, and goods.

10.3.6 Basic Requirements to Foster Growth and Stability

In order to pave the way for innovative projects such as Alp.Lab and, at the same time, not lose sight of the public mandate, the right to mobility for all, the ethical guidelines of trustworthy co-existence and respect amongst each other, we have listed the core evaluation criteria that are relevant to us:

- The right to mobility requires public authorities to address innovation and technological progress in a holistic manner.
- A responsible and ethical handling of consumer and mobility data is required.
- In our opinion, the free of charge and unbiased use of consumer and mobility data will not last long and will raise questions about present business principles in behaviour economics.
- Test laboratories contribute to the gain of knowledge for regulations and approval-relevant access roads and routes.

- Let us consider data acquisition from test series and test cases as tradable goods.
- The public sector needs innovation in order to make regulations feasible.
- Test laboratories help urban and infrastructure planners to better perceive and assess structural and procedural measures on existing and planned routes.
- In order to make autonomous driving suitable for everyday use, it is necessary to cope with the complexity caused by the multitude and functionality of the technical components provisioned by distinct organisations.
- The acceptance and operational readiness of autonomous vehicles in the population should be made visible and accessible to the population. Innovators should demonstrate its impact on society, through which they will be able to justify the viability of test laboratories and the required budgets.

We recommend applying the list in the context of any innovation project and right from the start-up in the kick-off phase.

10.4 DoktorB's Checklist

Important Basic Considerations

- Are you interested in making your region and/or city available for autonomous driving?
- In your opinion, what other tests other than those described above do you need?
- In your opinion, what does it take to convince the population of the advantages of autonomous driving?
- What criteria would you use in the future to select the use of your own vehicle in comparison to the use of autonomous vehicles made available?
- There are plans to organise postal mail deliveries with autonomous vehicles—what advantages do you see and what challenges would you project to those involved in postal services?
- You want to position yourself as a mobility provider in the future—with collaborations such as the Alp.Lab project in the foreground: how do you ensure that your offer is made visible in the vehicle?
- The BIM catalogue offers a wide range of solutions and mobility elements—where do you find yourself most likely with your company or start-up offering?

References

1. Alp.Lab (2018) Company website. http://www.alp-lab.at/. Accessed 18 May 2018
2. Bmvit – Federal Ministry of Transport, Innovation and Technology (2016) Automated – Networked – Mobile: Action Plan Automated Driving. bmvit, Wiesbaden
3. AC Styria (2017) Website. http://www.acstyria.com/de/index.php. Accessed 18.5.2018

4. AC Styria (2017) Fields of competence. http://www.acstyria.com/kompetenzfelder/. Accessed 18 May 2018
5. TU Graz (2017) Self-propelled cars – Styria becomes Europe's most diverse test environment. https://www.tugraz.at/tu-graz/services/news-stories/tu-graz-news/einzelansicht/article/selbstfahrende-autos-europas-vielfaeltigste-testumgebung-entsteht-in-der-steiermark/. Accessed 18 May 2018
6. Reference Styrian Mobility (2017) The future of driving. http://www.v2c2.at/fileadmin/user_upload/marketing/Presse_Clippings/2017-07-01_Steirische-Mobilitaet_-_ALP.Lab_Testzentrum_Autonomes_Fahren.pdf. Accessed 18 May 2018
7. ASIFiNAG (2017) Safe drive on 2,200 kilometres of motorways and expressways. https://www.asfinag.at/ueber-uns/. Accessed 18 May 2018
8. Alp.Lab (2017) Company presentation ALP.Lab from 23.01.2017 – Non-confidential
9. WFB Wirtschaftsförderung Bremen GmbH (2017) Driverless cars of the future: Bremen scientists work with us. https://www.wfb-bremen.de/de/page/stories/automotive-bremen/fahrerlose-autos-der-zukunft-bremer-wissenschaftler-tuefteln-mit. Accessed 18.5.2018
10. Netzwoche (2017) Autonomous Postbus will soon be leaving for Sion railway station. http://www.netzwoche.ch/news/2017-10-18/autonomes-postauto-faehrt-bald-bis-zum-bahnhof-sion. Accessed 18 May 2018
11. Volksblatt.li (2017) Postbus tests buses without chauffeur in Graubünden. https://www.volksblatt.li/nachricht.aspx?p1=region&id=167468&src=vb. Accessed 18 May 2018
12. WIRED (2017) South Korea is building an artificial city for autonomous cars. https://www.wired.de/collection/tech/k-city-suedkorea-testgelaende-selbstfahrende-autonome-autos. Accessed 18 May 2018
13. M-City & University of Michigan (2017) Leading the transformation to connected and automated vehicles. https://mcity.umich.edu/. Accessed 18 May 2018
14. gtm – Green Tech Media (2018) Report: Tesla ranks last for automated driving. https://www.greentechmedia.com/articles/read/report-ranks-tesla-last-for-automated-driving#gs.M_oPaOY. Accessed 18 May 2018
15. Flügge B (ed) (2016) Smart mobility – trends, concepts, best practices for intelligent mobility. Springer Vieweg, Wiesbaden

Always at Your Service! The Mobile Assistant 11

DoktorB Barbara Flügge

Abstract

I remember my past shopping experiences in the USA very well because of the one key service that Saks Fifth Avenue and Neiman Marcus offered. They provide their customers with a personal sales representative on request. The sales coach reads every wish from my lips, intuitively knows which brand I might like, which trends I follow, and which colour combinations suit me and which I would not keep in my closet. The personal assistant shortens our walking distances and unnecessary waiting times in the store and online, ensures a smooth process from selection, reservation and invoicing to the delivery of the chosen collection to your home. In my understanding, our expectations of mobile experiences will be increasingly shaped by experiences such as those described above. Picturing the mobile assistant like the coach in the store, what's the job and service profile of such a mobile assistant? To what extent may they influence our selection processes? As part of my trend scouting tour, I discovered the ENTOURAGE [1] project.

11.1 Motivation

The number of Smart Mobility projects has grown strongly in recent months and years. Meanwhile, project initiators do not only include sponsors, cities, and private companies. More communities and interest groups also become involved in the study, design, and conception of intelligent solutions.

D. B. Flügge (✉)
digital value creators (DVC), Bottighofen, Switzerland
e-mail: barbara.fluegge@dvcconsult.com

B. Flügge (ed.), *Smart Mobility in Practice: Good Car, Bad Car, No Car – Is this the End of Nurturing Our Mobility DNA?*,
https://doi.org/10.1007/978-3-658-43171-6_11

As part of my trend scouting tour at New Mobility World 2017, I noticed the ENTOURAGE [1] project here. ENTOURAGE—Enabling Trusted Ubiquitous Assistance—positions itself with the following statement: "The control of intelligent devices in the Smart Home, Connected Car or Intelligent Public Transport is ensured by closed platforms of the respective manufacturers. Manufacturers thus have an enormous influence on business models relating to intelligent devices. But they limit the possibilities of independent providers of innovative assistance services – not always to the advantage of users" (ibid.).

What is the unique selling proposition of ENTOURAGE? Officially it says "The ENTOURAGE project is creating an alternative based on an open ecosystem. It provides a novel architecture that enables direct interaction and integration of devices and services from different application fields. Legal security, data protection and an open marketplace for assistants, data and services are essential project contents" (ibid.).

Amongst the project partners are ENX Bosch, CONWEAVER, HaCon, the Fraunhofer IAO Institute, the University of Kassel and the Technical University of Darmstadt [2].

Despite the number of academic partners, ENTOURAGE presents itself not only as a research project. Rather, in addition to the investigation and project planning of technical issues, the focus is on the development and testing of a holistic approach: the ubiquitous digital assistant. This is intended to give consumers access to all those service offerings that can be of relevance in the life cycle of consuming.

Smart Mobility can be well explained in the ENTOURAGE project by the fact that we imagine the car acting as a "proxy" for shopping coaches, carsharing providers, and opportunity and office assistants. Services can also be presented at other locations, an example being a service kiosk, which is set up at the hotspots of a city or village and which we access through our citizen ID or dedicated user accounts. Think of a check-in terminal at the airport for example.

What is the ideal scenario for us from a consumer perspective? I imagine it this way: I get access to the offers through the digital companion. This is my omnipresent advisor, butler, chauffeur, shopping coach, and messenger all in one. For example, the digital companion informs me about trends, sports equipment offers and eases the registry to participate in golf tournaments in the near vicinity. The digital companion is indispensable and takes off the administrative burden, coordinating and preparing the meetings with my clients, organises logistics, and manages the vaccination checks before I am hopping off to the weekend. My mobile assistant helps me onsite, colouring my physical pick-up experience, so that I have time, i.e. at the store to have a chat with the in house sales representative. My preferences drive the roadmap and the insights that are being compiled in a reader's digest format. I will be arriving relaxed and well prepared at the point of destination. Meanwhile all items are placed in the shopping bags and ready to be picked up.

11.2 The Digital Companion: Introducing myEntourage

Today, there are few suitable digital companions that are good enough to support business travels and medical tourism ([3], p. 101, 161). The presence of a digital companion profited not only those who could no longer manage the way to the door without support—even those who are confronted with a multitude of offers and decisions due to illness or events were supported by a digital companion. If it were there!

The ENTOURAGE project targets a platform-independent solution. How can this be assessed from the current market situation? Let's take a look at the consumer market. Today, market researchers are already measuring the chances of success of assistance systems against the offerings of Google, Apple, and Amazon [4]. Taking a holistic view of the market development for mobile assistants we need to integrate data traction and third party applications, for example from the Smart Home and Medical Telediagnostics sectors, from OEM-dependent infotainment systems, and from chatbot-created unsolicited recommendations that derive from browsing and clicking analytics. If the consumer does not want to invest in Alexa or extra hardware solutions, we access the mobile assistant via mobile phone and/or a navigation system. It is to be expected that these are less comfortable to use at home and that certain functions such as the digital chaining of two to three "Alexas" across several rooms and spaces can only be mastered with more effort. But this effort is exactly what ENTOURAGE is trying to tackle.

The current offerings, such as Amazon's, have emerged retrospectively from entrepreneurial foresight and curiosity, the implementation of creative thinking, and as the result of a controlled innovation process. Therefore, it is not surprising that consumers encounter vendor-owned platforms. A neutral instance is less likely to be found. If necessary, it blooms in secret in isolated markets. ENTOURAGE is impressive in its ability to save users from being dependent on individual platforms and instead to offer unhindered access to assistants and thus to offers and modules—regardless of the platform originally used for this purpose. ENTOURAGE's goal is therefore not to position itself as another vendor-owned, central platform provider.

It takes little imagination to regard the platform of a single or beneficial provider as a bottleneck and gatekeeper. The more successful the provider is at offering additional services and value-added services, a lock-in effect emerges, and the greater the consumer's dependency on this provider. The consumer remains in the existing construct despite third-party offers from other platforms are more affordable. The dependency grows even further the more a platform provider manages to place those targeted services in advance, to take the consumer into the next life stage or consumption stage. The more skilful analytical cycles are used, the more precisely spatial decisions can be controlled in advance, taking into account patterns of movement, shopping, and meeting behaviour. Spatial decisions imply, for example, the move to a more suitable place to work, learn, and live in a more livable urban district. The more cleverly the decisions are digitally presented by the city administration, the more likely that a consumer would not necessarily perceive any influence in this respect for the time being, but rather decision support and comfort!

Digital companions such as ENTOURAGE require far-reaching integration of smart home solutions in order to integrate activities at home into a family's mobility plan, for example. Through the survey of suitable multi-modal solutions and the participation of vehicles or mobile modules, the test fields should be aligned as closely as possible to (a) the requirements of an intermodal mobility service and (b) using existing solution modules of project partners—as in the case of ENTOURAGE—and market participants.

Critical Assessment and Outlook

The extent to which platform plans such as ENTOURAGE's will prevail depends on the ability of established, albeit isolated, service providers to bind their users. The lock-in effect means that brand-loyal users cannot be persuaded to switch to another brand. It is also necessary to observe how OEMs and car manufacturers stand by the demand for platform neutrality. A further question will be to what extent consumers' car and OEM brand loyalty will not transform themselves into digital companion platform providers with an offering of an all-encompassing fulfilment of demand—beyond the car. Aren't cars to a certain extent already offering additional service packages for the drivers, co-drivers, and those that wait to kick off their mobility trip? The picture of in-car shopping comes to my mind—the car becoming the supermarket, premium boutique and shop window all in one.

11.3 From New and Established Players

The Intermediary as Trusted Advisor

ENX [5] has been an association of automobile manufacturers, suppliers and four national automobile associations since it was founded as the ENX Association in 2000. One of the main objectives of ENX is to initiate, shape and, if necessary, intervene and de-escalate its association structure as a mediator by establishing common standards and interoperable services based on these. ENX [5] describes its portfolio as follows:

> ENX is the joint solution of the European automotive industry for the secure exchange of critical development, purchasing and production control data. The cross-company ENX Managed Security Services and the service levels of the underlying network offer the highest quality and security features. At the same time, the network is as open and flexible for the circle of registered and connected users as the public Internet. [...]
>
> Founded in 2000, the ENX Association, a legally independent association of companies and associations Audi, BMW, Bosch, Continental, Daimler, DGA, Ford, Magna, PSA Peugeot Citroën, Renault, Volkswagen as well as the national associations ANFAC (Spain), GALIA (France), SMMT (UK) and VDA (Germany), monitors the service quality of its customers, operates central services of the ENX network and supports service providers in the efficient solution of emerging problems. (ibid.)

Another important role of the ENX Association lies in certification and accreditation [6]. This includes certified service providers, for example for the exchange of sensitive data in vehicle development.

In my view, ENX would not be able to survive as a supervisory body alone. This is where ENX's aspired role of becoming the *confidence anchor* is mission critical to its ability to become a market influencer. "ENX develops the application scenario Multimodal Mobility and investigates the significance of a trust anchor for the ecosystem from the point of view of the automotive industry". [1]

Even if solutions continue to be launched on the market independently of one another, ENTOURAGE's claim is that the resulting complexity should be solved in the back office. Consumers should not be made aware of the complexity or the role of the intervener or problem solver.

Projected onto the ENTOURAGE vision, I believe that the exchange of demand and offers needs to take place via pre-certified data structures, independent of the proprietary systems used by the providers—whether this is a shopping platform or a transport service. The stockpiling, marketing and operation of assistance systems, and digital companions will not change much for the providers in the market.

The Vehicle Takes a Back Seat

Let's look at a two-phase scenario driven by our taste and needs. In this scenario, we receive an offer to book a commercial delivery vehicle for just a few hours. We decide to take advantage of it because we have just renovated our study room and bought a desk lamp, an office chair and a mobile container in a new shop in the new city district. We opted to go to this shop in order to benefit from the onsite shopping experience. Imagine, the forecast for tomorrow is for sunshine and we therefore would preferably like to enjoy an excursion in a convertible instead. To get to the pick-up station, we take the subway and then an e-bike. We do not take an autonomous vehicle, as we want to exercise and follow health advice from our fitbit wristband. Convertible and commercial delivery vehicles are from two distinct brands, aka OEMs. Brands are not important to us and we have not been interested in which car brand we might get when booking our mobility experience in this case. The conscious booking of a brand-neutral option in no way diminishes the requirement fulfilment: the error-free and risk-free coordination of vehicles, regardless of whether we ask for mobility services in a static (shop) or moving (bus, car, bike lane) space.

If we had booked a brand-driven experience, our expectations then synchronise with the brand values, for example that a technology-driven brand like Audi ensures that the technical coordination functions and harmonises perfectly.

To be honest, I would like to see the preferences for interior design and entertainment transferred from one vehicle to another, too. Even if there are several days between the periods of use. In the example described above, our settings for information and entertainment would be automatically transferred from the commercial vehicle into the convertible. DoktorB's mood controller then welcomes us with our preferred genres of information, colours, road suggestions, and music titles.

Once myENTOURAGE is available, everyone—according to the vision—can buy a digital companion and use the corresponding assistance services.

Let's go one step further. It is not only human beings who order assistants such as myEntourage. Smart items, too, could book their personal mobility companion to consume products and services. App providers, platform providers and other assistance providers can get involved and join the service marketplace—as service providers and service consumers.

Who Benefits from the Digital Interplay?

The above-mentioned aspect of the neutral switchboard soon raises the question of who earns what from the digital companion and how? What does the business model look like? Traditionally, those platforms that exist differ in their business models, including whether the sign-up fee and subscription fee operate as per service, per period, or per individual functions and features. Despite their different origins, features and functions will be combined with each other or exchanged later. Regardless of who developed and offer features and functions such as audio control, landmark-driven navigation, and the graphical user interface, we, as consumers, expect that these features match digitally and convert into one seamless and well operating offering.

This digital plug & play challenges per se the architects of digital companions: from the design of the digital link, data exchange, notifications across the process module chain and up to the graphical user interface, the user must be accompanied consistently, regardless of where the offers and functions primarily come from.

In a completely open and self-steered ecosystem, this role of a trust anchor does not exist. How could the scalability of ENTOURAGE and other digital companions evolve in this direction? Is there not the risk that the control process and the convergence of data protection requirements will limit the number of tolerated service providers? As a mediator, ENTOURAGE would be able to demand chargeable certifications and accreditations, for example, as an entry point to work with ENX. If ENTOURAGE wants to move forward, it can take non-project related pilots and test environments to derive the terms of use!

With respect to the Smart Mobility BIM model (see Sect. 10.2), ENTOURAGE would match with the strategic and programmatic elements, as well as functional and technical elements. The latter are assigned to the four levels of Smart Services, Smart Data, Smart Products and Smart Spaces (◉ Fig. 11.1).

The BIM development plan comprises all relevant elements and supports project teams and stakeholders who want to offer products and services beyond mobility through its ecosystem-based approach.

More Profit Through New Roles

With ENTOURAGE in mind, let's model the position of Bosch in a smart mobility ecosystem. On the one hand, Bosch is a supplier of household appliances and thus fulfils consumer needs in the living domain. On the other hand, there is a further consumer demand created while travelling and not necessarily at home. The fulfilment of those needs can be met in a more targeted and timely manner. However, the close accompaniment of a user via any kind of digital companion requires an economic and market-compatible strength in order to actually be competitive. Could Bosch be my future digital companion? Would this then lead me to my favourite shops, depending on what I would like to prepare at home and the baking ingredients I need for my favourite cake? And would ENTOURAGE be able to read my guests' favourite dishes from their lips and help me prepare dinner? I could well imagine that the ingredients would automatically come home to me, that suitable recipe suggestions would be selected for me—measured by the

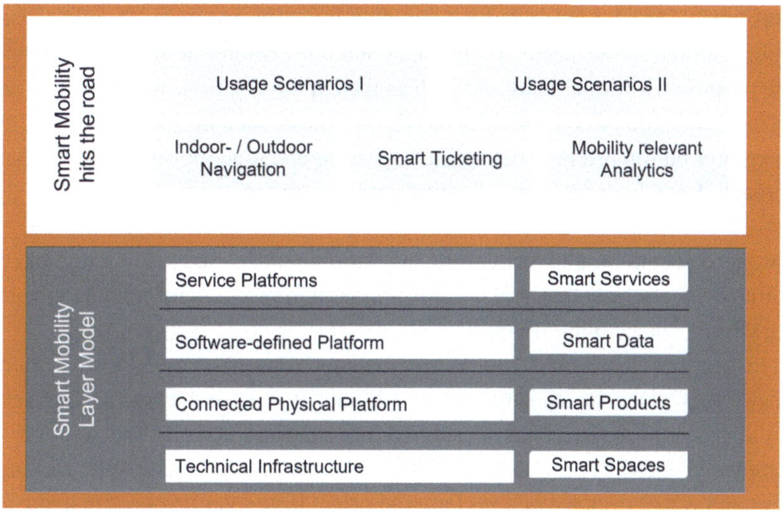

Fig. 11.1 The Smart Mobility layer model

degree of difficulty I could handle in the kitchen—and even more. There are no limits to ideating the interconnected and well-served scenario.

Once successfully deployed, ENTOURAGE could close the gap that other projects have left in the past. This gap is still prevalent not because the projects were not target-oriented or because they were strategically misplaced, but because the design and purpose of a service marketplace is neither B2B or B2C or C2B, it is a full and encompassing cyclical service mandate: B2B2C.

The operation of online and offline demand and supply functions is complex. I remember well my presentation in 2012 at informare! titled "Cloud as a distribution channel for web-based information services". At that point in time, my presentation raised eyebrows and question marks. Thriving through a holistic view and encompassing the interests of a citizen in a city, the digital data distribution amongst platforms should be becoming the mission critical ingredient to successfully serving our clients in any situation and space. This works even up to the point where the consumer leaves the proprietary shopping mall with a market-oriented portfolio of information, data, product and processing services and is then able to build his own shopping mall composed of own and third party offers completely digitally!

11.4 Well Advised with or Without a Car?

The role of the car is more than ever a discussion topic on news channels, at shareholder meetings, and in strategic board meetings. I will state the following: the car will become even more important due to the flexibility we now demand in order to visit, explore and operate in new geographical areas. Due to the compressed presentation of data in a closed space, this space—the car—demands all our attention. A prerequisite, of course, is a tailor-made offer: tailored to our needs, preferences and our explorer archetype. In terms of the latter, we propose additional features, such as the explorer button, which will be added to the vehicle—analogous to the mood controller we introduced in Sect. 4.3.

The explorer button offers you, as vehicle passengers, gradations similar to a gamification approach. Let's look into the following explorer options:

1. You want to get out of, and away from, everything you're used to? Our digital companion offers new contacts that you will meet in certain places. The vehicle will steer you there. Mobile geocaching, if you like.
2. A lot to discover? The vehicle opens up moments of surprise in unknown terrain. If the journey takes place in an autonomous vehicle, it does not require your selections regarding route preference, nor does it require your consent. Is it about a sports or event experience you don't know about yet? No problem: you receive tickets, parking reservations and a VIP package at a preferential price. If you get involved in a musical or a band, you will also get the latest music video presented in the vehicle to get you in the mood.
3. What's the news? The digital companion offers you news from familiar neighbourhoods, spaces, and cities. Your favourite boutique launches an advance sale—let's go! You've always wanted a new handbag—especially on Mondays—to lift your spirits. You can get them delivered in-house or, even better, in-car.

If we allow it, digitised offers can take advantage of our mood, past experiences, and purchases. It is us, the users, who focus on the car, whether this is because of its importance as a status symbol or because of its importance for mobility. Or because of the role we give him. Or because the car or its services adapt to our demands depending on the situation and offer us a second home or a second office. With respect to the latter, it is the car that bridges waiting times and offers lingering time. It evolves into a device analogous with the mobile phone and the tablet.

Compared to other devices, the vehicle provides the security that assistance is provided to the driver and front passenger while driving in the car. With the appropriate assistant, this sense of security is increased for further activities, literally when we use the car for work, shopping, or chatting! The situational demand we have on the car is increasingly fulfilled by the mechanisms that are already in the vehicle or can be fed in from the outside. With the growing demand for comfort and safety, the demand on the car to anticipate our preferences and needs will grow even stronger.

If platform approaches remain neutral and are not contractually linked to specific brands or suppliers, we as consumers continue to have freedom of choice. This means that we are not dependent on the car as the sole service provider. Imagine that you, otherwise, would need to sit in the car to benefit from these offerings. Luckily, we have further consumption platforms such as the smartphone, tablets, wearable, or other types of devices, all of which are still currently being worked on.

Under Chap. 12, we take a look at the everyday life of the Frohsinn family in the digital, autonomous age. Under Chap. 13, we take a look at mobile habitats and newer coaches. Not only do gamification, shopping, and fun play a role, but so do services in the field of healthcare, mental fitness, risk anticipation, and safety.

Consumers in Digital Driver's Seat

Digital companions help us to create and consume experience spaces inside and outside of the car. The situational demand we have on the car is increasingly fulfilled by the mechanisms that are already in the vehicle or can still be installed. Through brand-independent service platforms, the self-determined selection of services and the self-determined handling of one's own data, the consumer is in the digital driver's seat! However, with the growing demand for comfort and safety, the demand on the car to anticipate our preferences and needs and control them sooner or later will grow even stronger.

11.5 DoktorB's Checklist

Important Basic Considerations

- What skills and competences should your digital companion have?
- What ethical standards should your digital companion comply with?
- Which and how many apps do you use today to get an experience that is tailored to your needs?
- To what extent might your future digital companion apply your wishes, inquiries, and contacts?
- How many apps does your company feed its service offering into?
- Are you still pulling the strings or are there big, international players, who are outperforming you in prospect meetings and deal closing? If you look at the innovation process in your company from the customer's point of view, to what extent do you consider customer requirements and wishes?
- Will the car become your new sales and service channel?
- Does the vehicle play a role in your strategic considerations?

References

1. ENTOURAGE (2017) Project website. http://www.entourage-projekt.de/. Accessed 18 May 2018
2. ENTOURAGE (2015) Presentation ENTOURAGE – enabling trusted ubiquitous assistance. http://entourage-projekt.de/home.html?file=files/downloads/ENTOURAGE_Pra%CC%88sentation.pdf. Accessed 18 May 2018

3. Flügge B (ed) (2016) Smart mobility – trends, concepts, best practices for intelligent mobility. Springer Vieweg, Wiesbaden
4. Matta R (2018) Voice Control: Alexa, Cortana, Siri and Co. are on the advance, in notebookcheck.com. https://www.notebookcheck.com/Sprachsteuerung-Alexa-Cortana-Siri-und-Co-sind-auf-dem-Vormarsch.287744.0.html. Accessed 18 May 2018
5. ENX (2017) Company website. http://www.enxo.com/lang/de/. Accessed 18 May 2018
6. ENXO.com (2017) About ENX. http://www.enxo.com/_files/pdf/2010-06-22%20EoI%20Press%20Final%20EN.pdf. Accessed 18 May 2018

Autonomously Driving: Catch the Car, Get Our Data

12

DoktorB Barbara Flügge and Mark Zannoni

Abstract

We have learned in recent years about the how our vehicles collect our data for us—whether this is in order to provide the Connected Drive offer from BMW, or the mobile monitoring of our Daimler or Renault—our car has long been a data collector, data generator and data supplier for the automotive industry. But how does it serve your cities and your constituents? The interconnected car could serve very well the local behavioural economy and help cities design unique service offerings onsite. Technology-wise, the car doesn't have to climb all the levels 3, 4, or 5 of autonomous driving. You could gain on the spot mission critical insights that support infrastructure and asset management decisions. In this chapter, you will learn more about family Frohsinn and how their family lifestyle transforms.

We have learned in recent years about the data acquisition that takes place via vehicles—whether this is in aid of the Connected Drive offer from BMW, or the mobile monitoring of our Mercedes-Benz or Renault. Our car has long been a data collector, data generator and data supplier. But who does it actually supply to? Who benefits now and who doesn't

---

The transformative power through autonomous vehicles.

D. B. Flügge (✉)
digital value creators (DVC), Bottighofen, Switzerland
e-mail: barbara.fluegge@dvcconsult.com

M. Zannoni
IDC Corporate USA, Cleveland, OH, USA
e-mail: mzannoni@idc.com

B. Flügge (ed.), *Smart Mobility in Practice: Good Car, Bad Car, No Car – Is this the End of Nurturing Our Mobility DNA?*,
https://doi.org/10.1007/978-3-658-43171-6_12

at all? Do we deserve to benefit from our obvious willingness to share our driving, braking and selection behaviour data?

As a data hub, the car is an essential adjusting screw in a clockwork of activities and processes that functions as safely and reliably as its physical Swiss counterpart. The networked car doesn't have to climb to levels 3, 4, or 5 of automated driving to let all those who spin the threads in the background know where we would like to go, and at what time and in what mood. What about the car as a data hub in concrete terms? And where are we as consumers? We will now look at these questions from two different perspectives.

1. Mark Zannoni, from the analyst group IDC, shows us the view of urban planners, which shows the transformative power of Smart Data from different perspectives. It shows what urban planners can rely on in the future from the use of autonomous vehicles. This will include a variety of data in real time, which—if appropriately evaluated and applied—has a significant positive impact on the quality of life and sustainability of city centres. It should not be overlooked that data collection to this extent requires a political basis and acceptance to allow the collection of real-time data and processing using comparative data, as well as additionally further processing. The substrate in demand will be behavioural, movement, and decision patterns of individuals in dedicated consumption and life phases.
2. The inhabitants of Johannes' city in Cape Town experience the world of Smart Data in their own personal way. We will dive into the experiences of DoktorB's role model family Frohsinn and think about how we would react in their place. Additionally, we will experience how much our daily routines are simplified, since for us many decisions will be able to settle themselves as if on their own, as if driving by 3.

12.1 Transforming the City Through Autonomous Vehicles

Mark Zannoni

Innovations in autonomous vehicles will transform mobility, and as such, impact other sectors as well, such as logistics companies, delivery systems, car dealerships, and banking and financial services. The autonomous vehicle technology will also have a profound impact on cities, from their physical form and development to municipal operations and enhanced efficiencies. ◉ Fig. 12.1 presents some of the broader benefits of vehicle automation in cities. This chapter will focus on the potential for autonomous vehicles to transform cities, from their current development and operating patterns to the potential of greater sustainability and efficiency, and additionally the potential that will enable an enhanced quality of life.

More Vibrant Cities from the Better Use of Land

Cities and parts of cities that have been developed in the past century have been designed around cars, from wide avenues and parking facilities to regulations on roadway design and

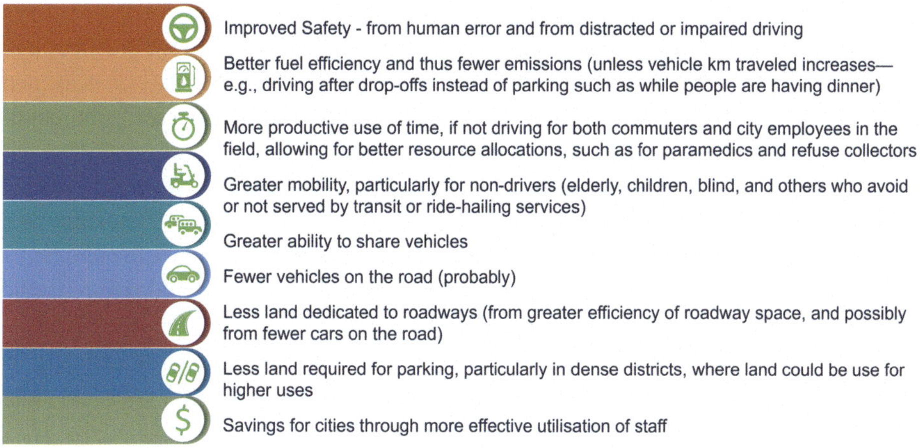

Fig. 12.1 The transformative power of autonomous vehicles. (Courtesy of IDC)

parking minimums. Cars, along with major thoroughfares and limited access highways, enabled cities and urbanised areas to sprawl outward, but this has resulted in the inefficient use of land, for both low-density housing and other uses. Autonomous vehicles provide an opportunity for cities to undertake a dramatic shift in their physical layout regarding existing and future land uses, and hopefully curb the pressures for a costly and unsightly urban sprawl.

Autonomous vehicles, ultimately operating as a "system" of traffic, allow for a more efficient use of roadways. For example, when a traffic signal turns green (though some believe the need for signalisation will disappear completely), all vehicles can begin to move without each car waiting for the one in front of it to move and automatically allow for a safe operating distance. Furthermore, vehicles could operate more closely to one another, both laterally and in following the vehicle ahead. Such efficiency assumes all vehicles on the road are communications-connected and autonomous, where precision and response-times are not affected by human limitations, vehicle maintenance is undertaken and all operating systems (e.g. braking systems) are in proper working order.

In addition to this operating efficiency on roadways, other contributing factors will reduce the amount of roadway required. These include social and economic changes, such as the growing trend of individuals foregoing private ownership of vehicles and instead relying heavily on ride hailing services like Uber, Taxify, or Grab, and the existing trend of work culture, such as working from home, more flexible working hours in offices, and the geographic dispersion of work locations which mitigate traffic crush loads as demand is spread over a longer period of time. In turn, this also enables carsharing to be feasible, especially if vehicles are autonomous. As roads and transport systems are designed for periods of peak demand, levelling, and reducing peaks helps minimise the size and number of roadways needed in a community.

Of course, these trends must be balanced against population and employment growth in cities. Furthermore, it has finally been accepted in policy circles that adding lanes to roads and highways does not, in general, alleviate traffic congestion; rather, they contribute to it.

Hence, the confluence of these factors results in a reduced need for roadway space in cities. As such, new roads could be designed more narrowly than current practices and existing streets may be reduced, allowing for bike lanes, wider sidewalks, or dedicated transit rights of way. Narrower streets create more walkable, more "liveable" thoroughfares, and thus establish a more pleasant urban community and experience.

Furthermore, the reduced need for road widenings and new roads, balancing any sprawl tendencies of a city, will reduce the capital expenses of building such infrastructure and the operating costs of maintaining them. This will enable savings in public expenditures or the possibility of using such funds for mass transit projects, which would reduce traffic congestion further while also improving urban quality of life.

In addition to reduced road capacity requirements and fewer cars per capita, it is expected that the demand for parking and the amount of space needed per car for parking will decrease. Accordingly, an increased amount of land in cities, particularly in busy urban districts, can be converted for other uses that will support a higher quality of life. These include public parks, which improve public health and air quality, or additional economic or residential activity, such as offices, hotels, or housing; this greater density of activity will support more shops, restaurants, and theatres, or other leisure or enrichment places, therefore further improving the neighbourhood and quality of life.

The reduction in land dedicated to parking is a result of two aspects of autonomous vehicles. Firstly, a vehicle could drop off a passenger in a busy urban district and then depart, either to transport another passenger or to park in a remote, less central location. Secondly, for vehicles that still park in urban centres, autonomous vehicles could park more precisely than a human driver and not require space for a driver (or other passengers) to exit the vehicle after parking. Thus, vehicles could be parked with only a few centimetres between them, thus maximising the space dedicated to parking. As an example, a garage that has a current capacity of 50 cars could hold 70 cars of the same size, and even more if cars are parked such that one vehicle blocks another after parking.

Many cities have "parking minimums", requiring a certain number of spaces to be constructed based on the square footage of a particular use built, such as residential, office, or retail. With autonomous vehicles and the reduced need for parking, such laws could be eliminated or at least significantly modified. Upon doing so, more compelling real estate projects can be expected as capital could be dedicated to the building or project itself without some portion siphoned away for parking, and thus contributing to the quality, scale, and viability of real estate projects. Larger projects providing more housing units or more office space given the savings from parking, for example, would further enrich the city, through density and the resulting increased support for local businesses, such as shops, services, and restaurants.

City Operations and Increased Efficiency

In addition to the physical form of cities, the autonomous vehicle provides the opportunity to positively impact city operations and related expenses. While vehicle autonomy will likely touch every city department, only a few examples of these benefits are covered here.

Approximately 95% of all traffic accidents are the result of human error. With autonomous vehicles, the number of accidents should decrease dramatically. As such, fewer city police hours would be spent responding to accidents, writing reports, and directing traffic for congestion caused by an incident. These hours saved could be translated into fewer police on a municipal payroll, and thus a smaller budget would be required, or the ability to reassign such officers to other police work, such as community policing or crime investigations. Fewer road accidents also mean fewer emergency medical and emergency fire department calls, thus increasing savings or allowing the re-assignment of tasks in those departments as well. Additionally, hospital emergency rooms would be less crowded and, if staffed at the same levels, faster processing and treatment of other emergency medical conditions visiting the emergency room would result.

On the topic of policing, it should also be noted that cities that rely strongly on revenue from traffic and parking infractions as part of their budget should be prepared for decreased levels on the assumption that the autonomous vehicles will not violate traffic laws. Correspondingly, this means that police who are dedicated to catching traffic offenders could be reassigned to other police duties, as mentioned above.

An autonomous ambulance would allow emergency medical technicians and paramedics to be dedicated to patient care in the back of an ambulance instead of driving. Though usually working in pairs, in cases of disasters or where personnel is spread thin from an incident site, one paramedic could stay with a patient, while another paramedic could care for another patient in a separate ambulance going to the same or another hospital simultaneously.

Autonomous refuse trucks (currently being tested in Sweden) allow for more efficient refuse collection as all personnel with the truck in the field can be loading the truck, instead of one person either driving the truck or continuously entering and alighting the truck as it moves along its route, which can lead to occupational health issues. In addition to optimised vehicle operations (which saves fuel and maintenance), the trucks provide greater safety using sensors and object detection and can operate in close quarters.

In addition to autonomous vehicles on city streets, autonomous commercial planes can be expected, at least in some countries. Airports, whether owned by cities or other authorities or companies, exist almost always to serve a city and, as such, play a critical role in a community's livelihood and economic prosperity. As such, they are an integral part of a city's transportation system. Aside from changes to tower and ground operations at airports, with the introduction of autonomous planes in commercial service, cities may be able to obtain new routes and destinations, growing their economies, as airline economics shift, benefiting from airline savings in labour costs.

From steady acceleration and other vehicle operations capabilities afforded by autonomous vehicles, fuel efficiency will improve, meaning less fuel consumption and improved air quality. (It is expected that autonomous vehicles will ultimately be electric-powered once battery innovations allow for longer operating times and faster and more ubiquitous charging capabilities.) Better air quality, in turn, will result in a healthier (and thus more productive) population, particularly for those susceptible to respiratory issues. Additionally, fewer emergency room and other medical visits for asthma and other air quality-related issues should reduce public health expenditures.

With autonomous vehicles, profound amounts of data will be collected, particularly as cities and vehicles add an increasing number of sensors and sensor types, and as sensors themselves become more sophisticated. The vast pool of data, which can be analysed in real time, and processed through algorithms and analytics, can then be used in profound ways to transform the city, such as identifying a traffic incident instantly and automatically dispatching the closest police officers and, based on the analysis, emergency medical personnel, while also warning other travellers, benefiting any injured parties as well as minimising traffic congestion issues that may otherwise ensue. Another benefit is the ability to produce real time reports of infrastructure conditions, including, for example, a new stress on a bridge or ice on a road, and the instantaneous reporting to, or control of, oncoming vehicles to avoid such areas. Additionally, data on travel behaviours, locations, routes, and times can be used to develop new and meaningful mass transit routes and transportation policies, such as congestion charging for certain areas or routes to affect behaviour and minimise congestion and traffic issues for all roadway users. The applications for public policy and improving the quality of life in urban centres through the use of newly available data on intracity travel and transportation system conditions, combined with direct communications with autonomous vehicles, will truly transform not only urban mobility, but life in cities as a whole.

Conclusion

Autonomous vehicles—across vehicle types from private cars to garbage trucks and even passenger aircraft—have the potential to profoundly transform cities in positive ways. Such technology will allow for more vibrant, compact cities with narrower streets and greater pedestrian activity, coupled with more land available for parks, housing, and commercial uses over unsightly parking facilities. The operational efficiency of cities could increase significantly while related public expenses could correspondingly decrease or be reallocated for increased quality of life activities. City police, for example, would have fewer traffic accidents and traffic violations in which to respond, which are time-consuming activities. Consequently, a city could reduce its spend on police or reassign officers to other meaningful activities such as community policing or crime investigations. Given the transformative impacts on land use and city operations that autonomous vehicles will bring to communities, local governments should begin contemplating immediately how best to position and leverage this transformation for its greatest potential benefit for the city.

12.2 Autonomous on the Road: Enjoy the Ride, Family Frohsinn

Barbara Flügge

How's the family Frohsinn doing now? Johannes, as we reported in Sect. 3.4, has successfully launched his mobility programme. In the first step he set up community pooling, based on level 1 and 2 autonomous driving and targeting car owners and taxi drivers willing to share their vehicles. The owners and taxi drivers agreed to offer a sharing service. Urban neighbourhoods and residential districts are well suited for this. Our team Mobilize4Data won the City Hack competition; the team will therefore be digitally capturing and accompanying the transformation towards autonomous driving. This project requires on-site test users who are willing to upgrade their vehicle using a digital step-by-step concept, apply this, and make their increasing knowledge about behavioural changes accessible to the research team. In addition, the team then utilises the findings and quality measures to quantify new shareholder value models.

Daniel and Diana, the couple Frohsinn, have two cars, one of which is leased, whilst the other is owned by the family. Both cars have a navigation system and certain entertainment functions as standard. Diana's car is also equipped with a comprehensive infotainment system. For her daily journey to work, a flower shop, she uses her car to travel from their home region to the city. Lovingly she calls the car "007"—always in use and reliable from A to B and C, just like the role model James Bond.

Daniel, Diana's husband, flirts with abandoning his car altogether. He likes to drive. His daily journey to work, to his meetings and appointments, and to the construction sites where his crew are on the road, however, increasingly annoys him. Fatigue creeps over him while driving. In the evenings, he is busy maintaining his administrative business tasks and his accounting work is becoming increasingly stressful. Whilst his company was growing and becoming more profitable, he had purchased a diesel vehicle. The increasing discussion about driving bans for combustion engines prompts him to think about an alternative because most of his clients reside in cities where he really can't fight his way around without a car! This, incidentally, is also the case for colleagues who, like Daniel, have started a solopreneur business, such as electricians, sanitary installation experts, and a direct marketing agency. Back to the car: the diesel vehicle holds many benefits because it is inexpensive to operate, can withstand high loads and, despite many kilometres on the speedometer, is still a reliable car.

The children, Paul and Paula, love to travel autonomously. This allows them to call up sports information while on the move, head for the shortest route to the nearest meeting point with their friends and get a glimpse of the latest movie releases. Of course, the information and services offered should meet their taste and what they would think sounds interesting. Being distracted along the route, time and distance from the home base to the city are easily bridged. 35 kilometres can quickly be prolonged into a torturous journey when inbound traffic towards the city centre causes traffic jams—for drivers and passengers alike. Isn't it wonderful how much Diana can rely on 007 so that her kids are taken care of safely, whilst also having fun?

We are now concerned with what happens to the car in terms of being the datahub, aka the data collector and platform all in one. Our first hypothesis is the following: the acceptance of in-car offerings for Paul and Paula, Diana, and co-travellers, work so well because neither Paul and Paula nor Diana are concerned about how much their behaviour controls the offer and how much profit can be earned by providing behavioural data. We call this the behavioural economics hypothesis (Thesis 1).

Our behaviour, as well as additional information we unwittingly share, seem to be a free of charge, always available asset. Neither Paul and Paula nor their parents know about the influence of their data on new advertising offers for young people who will be travelling alone in autonomous vehicles in the future. In the value chain of service designers up to service providers and brokers however, consumer data has a price tag. Furthermore, Paul, Paula, Diana, and friends also do not get to know about renumeration when data is being transferred from provider to broker and so on. If they knew, would they be concerned about what a condition-based business model might look like: they might be concerned how their personal information and behaviour could become a tradable asset for certain days, events, or seasons? If so, what should this information be worth? How much income per day, week or month could Diana expect to be compensated? And to what level could Paula and Paul increase their pocket money by consciously sharing and renting their data?

In the meantime, while Paula and Paul meet their friends, and Diana is busy in her flower shop, Daniel is ideating about how his work and private life would change with an autonomous vehicle. He's signing up for a smart mobility project with DoktorB and Johannes. The initiators in his town are deliberately looking for small businesses and entrepreneurs who would experience the biggest cuts and financial burdens if they turned away from the diesel car. Daniel meets Johannes, the mayor, and our Mobilize4Data team and depicts his vision—the vision of what it would be like to actually move autonomously—to his work assignments, to visit his crew at their construction sites and carry out safety checks. He would never have thought about what impact this might have on his customer relationship—until Johannes was looking for volunteer business owners. The entire benefit of autonomous driving for businesses was new to Daniel. He believed before that autonomous driving only served those who have enough money. He stated: "You can read about the A8s and S-classes that are packed with new technology in the online magazines". He became even more excited after having joined the project with Johannes, becoming part of the test and contributing his experiences. Read below what Daniel is envisioning.

Daniel Is Sharing His Vision of a New Everyday Life

- Daniel's scheduling of tasks is now part of the "drive" in his car—our team helped him with the synchronisation of customer master data, geo locations, and meeting schedules.
- The car takes Daniel to all his appointments and makes sure that traffic jams and any delays are taken into account, and his itinerary is adjusted to arrive on time.
- For the time being, he does not care about the ownership of such a car.
- While he is being served and moved, he looks through his workday checklists again, which he has created the previous day during day time—something he was able to do because of the quality time he has gained.

- In the car, all essential project information is transmitted and displayed onto the cockpit display—finally a reasonable size and not such a tiny smartphone screen! Daniel has always been annoyed that even a tablet's display is too small for his application areas. The displayed data informs him about the construction sites, where his team is located and where all his stakeholders are meeting in the remote onsite office container to go through the workload and signoffs.
- The checklists are presented in the car and on his table as swipe files. This allows him to store his notes directly and do office work he would otherwise have done in the office on site.
- He arrives totally relaxed to the prospect meeting! Daniel's always on the run. Paul and Paula are already happy in the morning, when the breakfast table is a bit quieter. After all, school is not always a pleasure. Dad shouldn't always think about what could go wrong, Paula thinks worrying about Daniel's temper.
- Johannes' team of students and DoktorB, the transformation and technology partner, also presented to Daniel and the other test crew members these great smart data glasses. In case he gets stuck, he utilises them to retrieve all necessary data about a construction site including the layout plan, the crew and the actual workplans. There is always a lot going on and he wants to present himself as well prepared in order to convince his new clients.
- Daniel deals with team absences every now and then. He has to anticipate what could happen in the case that a colleague is unavailable, which puts an additional workload and stress on the rest of the team. With a small team like his, the impact can't be easily mitigated. The bad news will appear on the display: Alfred has called in sick!
- Thanks to the foresighted planning tool, the car issues Alfred's absence notification to his tablet before kicking off his work trip.
- The car proposes two possible alternatives for the team to close Alfred's gap. Both are within geographical reach and available. In addition, the car offers Daniel a replacement colleague. Johannes had the idea that colleagues who do not work in one company could serve other companies. Meanwhile, this has made its rounds in the district. And two specialists who recently lost their jobs in a large company about 100 kilometres away appear on the Joker list. Joker list is the right term, Daniel resonates. They help me and get a chance to demonstrate their skills and build up further references.
- Daniel can think of so many other improvements.
- The test takes its course.
- At home Paul and Paula find a relaxed dad again, who finally has time to listen to them.
- And Diana? She is looking forward to spending an evening together with her husband—without work. Daniel invites her to a new restaurant. What about the glass of champagne they would like to enjoy? The car picks them up and brings them back safely.

You may think, beautiful new world, but it's not that simple! Even if the full suite of autonomy of vehicles were to be delayed, supporting and balancing functions are already being introduced and made available in cars ▶ Sect. 12.3. The technical advancements in the scenario described above is one side of the coin. What about Johannes and his project team? Isn't it great how he and his colleagues stand up for their city?

Our second hypothesis is therefore: the intelligent data processing of mobility consumption is becoming a key characteristic of a smart city. Herein, the secondary service providers could benefit from data collection within and around vehicles and passengers' behaviour. Only if the municipality government and stakeholders themselves are forward thinkers and take action can there be a return on investment for the city's citizens and constituents (Thesis 2). A key prerequisite is the willpower to change! Conscious testing and learning works alongside this drive as a means of proving the way power of government. This includes learning, for example, how to deal with intelligent asset management and intelligent interconnectedness. I imagine how the city turns into a test laboratory: the road surface is equipped with induction technology, the traffic light system connects with the street lamps and any other structural facilities of the city, such as buildings and bridges, are linked up, too. Sustainable energy control depending on urban consumption is carried out by taking into account the state of natural resources such as sunlight and water—utilising adequate and affordable sensor technology.

On the other hand, a Smart City staffs his business units and designs its role model to maximise the operational value of new means and technology. The strategic goal of leading a city sustainably, in a future-oriented manner, safely and inclusively is Johannes' mission. Together with his management team they define the how and what. Each city defines the rules and guidelines in accordance with its vision and mission—just like a corporate leadership would look like. There might be in the near future distinct different rules in Diana's living place and in the city she would visit over a weekend with her husband and children. The autonomy of cities results in different characteristics of data flow and business model offerings. It is the role of the platform provider—which is the car in our case—to determine what is activated in terms of the specific rules and guidelines upon entering the city limits.

We strongly recommend spending time on the role model and business modelling elements. The ethical sharing of data and resources in urban spaces is the basis for successful Smart City concepts. All stakeholders—citizens, businesses, and the administration—must be able to benefit directly, otherwise concepts will turn into content-free shells and will lack citizen participation. Sharing data not only provides intelligent solutions for street lighting and traffic management; the digital touchpoints are opening up entirely new ways for governments and businesses to communicate, interact with, and serve constituents, and ultimately help small businesses like Daniel's.

Thus, the opportunities offered to Diana and her family are directly related to the (1) degree of autonomy of urban decision-makers and the (2) degree of participation that these same decision-makers grant to citizens such as Diana, Daniel, Paul and Paula. Both hypotheses,

- Hypothesis 1 "Behaviour controls supply and demand" and
- Hypothesis 2 "Intelligent city meets open-minded decision-makers and users",

are critical success factors for sustainable onsite growth.

Speaking of sharing, the digital value share is many times higher than the physical value share. It is the digital values that grant a growth spurt to the physical, infrastructural space! This is due to the long-term implications of short- and medium-term decisions. An example: within three to five-year planning approval procedures, cities determine the twenty to thirty-year values of tram and underground lines, airport extensions and station construction, tunnels and intersection planning, as well as the construction, expansion and operation of event and sports buildings. Successful, growth-oriented cities do not seek to compare electoral periods at local and national level. Rather, they measure themselves against the profile of their city, which is made up of the profiles of the individual citizens, residing companies, education, and higher education spaces, and innovation and research institutions. We usually target around 14 identity creators in our projects.

It is in the hands of local public administrations in cities and districts to take advantage of autonomy. Even though the beginnings of autonomous driving have made their way with partial routes and regional focuses, it is already possible today to set the course in terms of structure and process organisation. The tasks presented in ◉ Fig. 12.2 serve as a starting point for a joint development of intelligently designed mobility at a location!

There are no limits to the variety of ideas to establish intelligent living spaces. Because we determine the limits!

In Johannes' city, Sophie and Karl also belong to the stakeholder group for the test project. Sophie is an employee of a company for the disabled. The company is struggling continuously to survive due to ineffective and diminishing sales. Sophie and her colleagues craft wooden toys. Very often they lack delivery personnel. Either learn or drive—most of their drivers are undergraduates and try to do both, driving and learning. If they could learn on the go and the driving could be complemented with extra mobility support, this would be something different: imagine that there is a driving learning and training laboratory!

Karl is a farmer and sells his fresh fruit and vegetables at the weekly market on Saturdays in St. Gallen, his local town, and on Wednesdays at the organic market in another region. Leftovers are taken back to this farm. We worked with farmers and bio supermarkets and asked them what an intelligent car could do for him and his colleagues. Daniel shared his autonomous driving experience. Karl himself participated in our interview series that we set up for the students at AMD Munich, the Academy for Fashion and Design. Initially he wondered if we advised him in redesigning aprons and rubber boots. Far from it! We set up the students project to survey food logistics, food waste, and what innovation could do to resolve organic and fresh logistics for small farmers.

Service Provider

- The range is diverse: classic consumer goods industry, automotive and retail and wholesale companies
- New providers through things and infrastructural systems using artificial intelligence and machine learning

Hosting service

- Hosting is provided by in-house or third-party providers such as IT companies and system integrators
- The same applies to established hosting providers such as Amazon Web Services, Google and Baidu.

Service Gateway

- In-house or third-party providers such as IT companies and system integrators
- OEM development units
- These ensure that supply and consumption channels are provided with standardized formats

Service Aggregator

- Bundling is carried out, for example, according to routes and driving sections in analogy to route control in air traffic
- Instead of offering vehicles, mobility packages are offered
- Decoupling of system and system operation
- Providers of autonomous vehicles act as aggregators on the market

Service Channel Maker

- Depending on the vehicle type, module type, location, routes, experience and event types, the preparation for sales takes place
- Other examples include brand ambassadors from the sports entertainment, consumer goods and industrial sectors
- There are also device providers and interoperability providers

Service Broker

- The autonomous vehicle or module can offer services independently
- Depending on the vehicle type, module type, location, routes, experience and event types, distribution takes place
- Brokerage takes place, for example, via mobility managers and service providers as well as physical and digital hotspots in cities and municipalities

Service Consumer

- Individuals
- Groups
- Earmarked bundling / grouping
- Cargo
- Means of transportation

Fig. 12.2 Autonomy guides offerings into a new, innovative service model

12.3 It Couldn't Be More Comfortable!

Barbara Flügge

From the Simplicity of the Decision

Let's get back into our own decision-making routine. How spontaneously have we been watching for a new car and checking the job market? How rigidly do we kick-start the process? For example, when you have been looking for a new plasma screen. Have you ever booked a two-hour test session to check out the perfect coffee machine and even paid for the entire trial session? Let's have a look further into the mobility element of the purchase. Have you considered whether you need a car onsite for the purchase and/or in-store collection of the TV screen? Maybe your thoughts go further and you envision yourself parking the car at the Park and Ride space and taking the bus to the TV store. Maybe you do not want to move your car—despite having one—and would rather book the home delivery service for an extra ten or twenty dollars. While your thoughts are travelling around the ideal process of getting your TV screen installed in your home cinema, you might wonder if you needed the car at all.

Depending on the purpose, the budget, the desired comfort, and the time available, there are many details to be clarified, inquired about, and called up in advance. It's no different for us when we travel. Passenger travel management is about designing mobility bundles that are measured against the customer's preferences of budget, comfort, and time. You might want to check out our use cases in (cf. Part II [1])

In order to reduce planning efforts and avoid unnecessary costs, we had introduced the concept of the digital concierge (cf. [1], p. 161 f.). Imagine that it has turned into your everywhere supportive digital companion, a suite of services made available to travellers and planners at all times. The digital data matchmaking technology will calculate the personal optimum of comfort, budget, and time—these three decision criteria we refer to as the decision triangle. Each travel segment has its own decision triangle for each of the travellers and it is the platform that sets all segments in context. Seamlessly, creatively, and logically choreographed, the end-to-end mobility service will meet your overall objective and subgoals.

Traversing on the roads of desire, with autonomous driving everything will be different! Instead of asking for extensive offers and freeing up our precious time to plan, replan, adjust, book, and rebook, we entrust ourselves to the autonomous vehicle. No matter whether we are served by a hyperloop for long distances or an autonomous passenger compartment or an autonomous minivan—the latter one might even remind us of our parents' newest camping bus in modular design [2].

Today we decide according to monetary, social, and planning aspects ◉ Fig. 12.3. We consider the following:

- When do we arrive?
- Do we have time for a dinner before the appointment?
- What do we do if the train is running late?

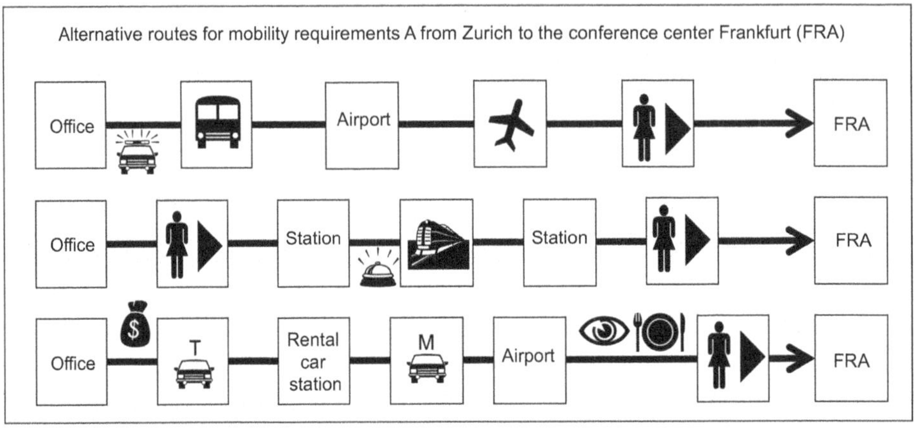

Fig. 12.3 Today we decide differently! Pimping up your business trip. (From [1]; courtesy of Springer Fachmedien Wiesbaden GmbH. All Rights Reserved)

- How do we handle delays?
- Whether there are any changes to the overall schedule and meeting deadlines?
- Where is the phone number of the LinkedIn contact we were supposed to meet after the business meeting?
- What are our options to do something for personal well-being e.g. walk the distance from the airport terminal to the conference centre?

Our scope for decision-making looks different in the mobility of tomorrow with autonomous driving! It is a matter of completely different considerations (◉ Fig. 12.4).

With mobility autonomy we as consumers notice a change:

- We have less effort in transferring between mobility means.
- We choose what is offered to us.
  - We get the option of whether we want to travel individually and autonomously or prefer group travel in order to make our contribution to sustainability.
  - We opt for the mobility providers offered to us instead of looking for alternatives ourselves. It is irrelevant to us which provider actually serves which transport segments and routes. We rely on the service provider and its offering.
  - We pick the mobility service provider who offers us a premium service which we would like to book consciously on the next journey and which we have not experienced yet from any other provider to date.
- We are prepared to accept that:
  - Isolated and separated travel segments are bundled up,
  - The end-to-end trip is a result of digitally calculated optimum
  - The mobility providers we booked to date no longer exist and will no longer be serving us.
- The Decision Triangle is depicted ◉ Fig. 12.5.

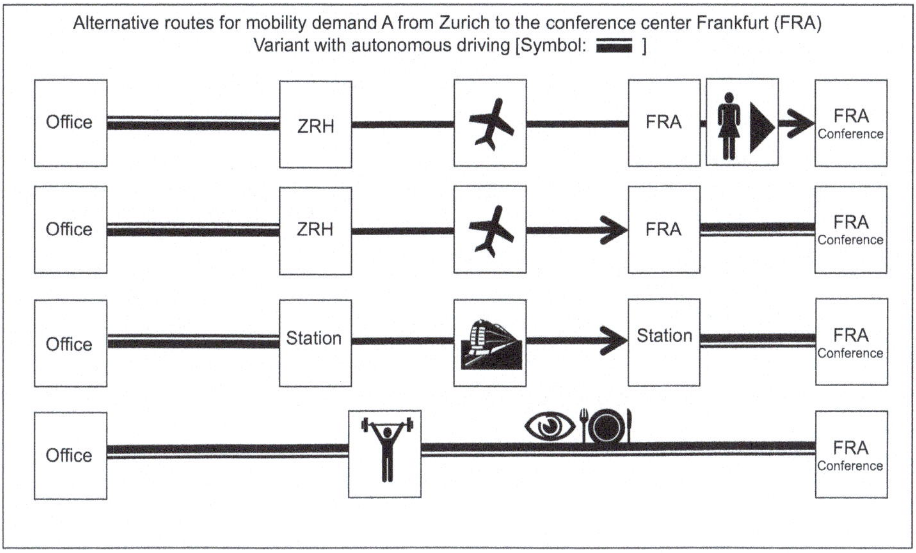

Fig. 12.4 Tomorrow, mobility concerns are not our worry at all!

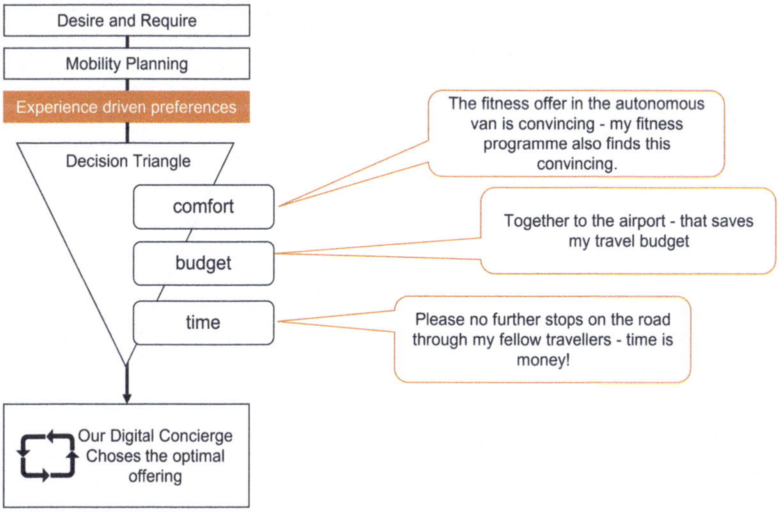

Fig. 12.5 Travellers in the hands of autonomous mobility

Diversity as Becoming a Core Asset of Intelligent Mobility Management
City-wide mobility managers will be truly busy. Additional options arise through an increasing mix of mobility means, whether this is a drone, autonomous mobility, or others (◉ Fig. 12.6).

Mobility is a driving force in our economy and therefore influences prosperity in cities and rural areas. As much as autonomous mobility on the streets has been seeking regulators' attention, any kind of moving modules as outlined in ◉ Fig. 12.6 are challenging existing governance guidelines and standards for safety and reliability.

There will be the need of airspace, ground space and intermediary space regulations and coordination. We picture a city control tower. Similar to the airport control tower, monitoring and protection of both ground and air space will take place. Mobility managers are also called upon to tender, award, or licence the rights to airspace for drone routes and self-driving/self-flying modules. Ideally—depending on the legal situation—this is done in clarification and agreement with the traffic rights regulation for *airspaces*. Once the operational concerns have been settled, it will become clear whether the number of mobility providers offered in a city should be reduced or increased. Irrespective of this, consumers thrive for their ideal experience. Experiences are the new currency. Herein, whoever does the set-up and maintenance of infrastructure moves backstage. Service providers and brokers receive the core attention.

What's the impact on intersections compared to the degree of expected excitement? Municipalities are forced to reiterate their mobility calculation, business, and service

Self-propelled vehicle	Grouped & Bundled self-propelled vehicles	Mobility modules & chambers self-propelled or remote propulsion	Coupled modules, cells, chambers self-propelled or remote propulsion	Cargo cells with remote actuator (left) and/or self-propelled (right)	Locomotion in the tube / Hyperloop	Mobility base in air and ground controlled territories (e.g. for drones) *)
Carriage of • Passengers • Cargo	Carriage of • Passengers • Cargo	Carriage of • Passengers • Cargo	Carriage of • Passengers • Cargo	Carriage of Cargo	Carriage of • Passengers • Cargo	Carriage of Passengers
Individually on the way	Sharing routes	Individually on the way	Sharing routes	Individually on the way	Sharing routes	Individually on the way
• Any routes • Routes by choice	• Any routes • Routes by choice	• Any routes (self-propelled) • Defined routes (remote propulsion)	• Any routes (self-propelled) • Defined routes (remote propulsion)	• Any routes (self-propelled) • Defined routes (remote propulsion)	Defined routes (remote propulsion)	Any route
• Individual travels • Shared travels	Joint travel	Individual routes and sites	Route and location optimisation	Route and location optimisation	• Individual use • Group use • Pre-determined routes solely • Intermodal routing feasible with pre-/post-coupling of 3rd party travel means	Individual travel feasible within authorized air space and/or ground space
	Flexible coupling options within route availability	Route optimisation is triggered by adventure targets	Any coupling options within route availability	Any coupling options within route availability	Pre-defined and selective coupling	Flexible coupling options
		Self-determined focus on interior and exterior while traveling	Versatile module use offering e.g. silent zones and networking zones			Observing the environment (on earth and in the air)
		Consuming services of individual modules	Purpose driven ride and service consumption in pre-defined compartments	Purpose driven modules serve e.g. cold supply chain, thermal logistics	Purpose driven ride and service consumption in pre-defined compartments	Individual use drives service needs

Please not that the focus of the category grid is on the travel means themselves, nevertheless all proposed travel means could / should become part of intermodality | *) alternative travel mean: spaceships and extraterrestial drones

Fig. 12.6 Diverse means serving smart cities and rural regions

model. The new variables ask for the minimum number of total trips, the number of intersecting travel paths, and the appropriate dwell time at intersections. In our concept "New Mobility for Rural Areas in Lower Saxony", we ideated travel journeys and hero journeys, and designed a digital data collection layout, for each of the personas we identified.

How big is the chance that a purchase or booking of a concert ticket will take place on the way? And how great is the chance that this will take place during the remaining stay? We as mobility consumers are particularly concerned with the bare essentials.

Old-fashioned waiting times make us wait for the completion of the booking of the means of transport or for the feedback of the *taxi operator*s or its app, wait for boarding, then also for the searching of the route to our destination, the payment of the driver, wait for the receipt to arrive in our e-mail inbox, before finally leaving the vehicle and unloading the luggage, then moving towards the next destination.

Adaptability of Man and Machine

With the experiences that result from autonomous driving, we inevitably adapt our expectations to the experience offered by the vehicle. If we change vehicles more often, we remain flexible and the appetite for unknown experiences is stimulated anew. If the digital assistant repeatedly offers us the same bundle of experiences, our demand behaviour also adapts. The curiosity for new experiences retreats and we remain in the moments that are offered to us.

The Coach for All Cases

The car is transforming into a living, learning, loving space; a space where we plan decisions and execute them. Soon we will need another term to describe the equipment of the car as a multi-purpose living space. Let us paraphrase this with the term *In-Car Living*. This has a negative aftertaste depending on the search engine result. In-Car Living often goes hand in hand with the population group that can only afford one car to live in and stay overnight. In a positive sense, In-Car Living is finding a new way of life: we live in a car and need everything we would like to have at home [3]. That's why the digital concierge doesn't just accompany us in the car. The digital concierge stands by us with advice and action in all our activities regardless of where we are and will be [4, 5].

In the case of Diana and Daniel, as well as Paul and Paula, ▶ we have already seen in Sect. 12.2 how the car develops into the BestBuddy. For health and fitness checks, Daniel uses functionalities such as regeneration, balance, and the check of his vital data. A digital health avatar suggests various exercises, ranging from small meditative contributions to relaxation exercises and back massages. Depending on the situation, supportive functions such as breaks in the fresh air are suggested.

Diana herself receives feedback on her driving behaviour, traffic information, weather changes and reservation options for parking spaces. She wants to keep the 007. To optimise her journey, she digitally passes on appointments and convenient meeting points along the way to her friends. So that her appointments do not become too monotonous, she constantly changes her preferences for locations. The family dog Charlie is often walked by the drone Harry. This allows Diana to devote herself to her leisure planning. And she can be creative to invent new flower arrangements. On the digital drawing board, she designs her creations and captures notes about the important preferences of her customers. In the shop, the input is already sent to the creative board and the colours and shapes, as well as the currently available flower varieties, are compared with the last orders of their regular customers.

Paul and Paula are more than happy. In the morning harmony finally arrives at the breakfast table. Charlie doesn't have to be walked by the kids early in the morning. The parents have finally gone out together the night before. There is quality time and good energy to exchange about cool ideas already at the breakfast table; Paul and Paula could improve their pocket money with BestBuddy. They have been publishing their tips for cool music titles. By making songs known on social networks, they receive coupons which they can then redeem again in their favourite stores. Since the whole thing is controlled by the click behaviour, a small amount is generated over time. They can use their pocket money in a targeted way to save for their next visit to the cinema, because they want to spend that together with their friends in a truly onsite and analogue manner.

Who's Driving Who or What?

As digitisation progresses, we will see an increase in offers being placed in digital companions and concierges. Their action parameter is growing way beyond mobility and living. So, let's step back and asks ourselves who is steering our decisions: Do we configure the digital companions or do the digital companions control and guide us?

12.4 DoktorB's Checklist

Important Basic Considerations

- Would you like to join Mobilize4Data? We are looking for interview and project partners for this exciting project. We conceptualise and execute Mobilize4Data service cities and their local economies; we evaluate the transformation and the value of autonomous and mobile living spaces for different target groups. Fill in the form https://www.barbara-fluegge.com/Mobilize4Data.

References

1. Flügge B (ed) (2016) Smart mobility – trends, concepts, best practices for intelligent mobility. Springer Vieweg, Wiesbaden

2. Matta R (2018) Voice control: Alexa, Cortana, Siri and Co. are on the advance, in notebookcheck.com. https://www.notebookcheck.com/Sprachsteuerung-Alexa-Cortana-Siri-und-Co-sind-auf-dem-Vormarsch.287744.0.html. Accessed 18 May 2018
3. The Tagesspiegel (2015) Connected driving – the third habitat. http://www.tagesspiegel.de/mobil/vernetztes-fahren-der-dritte-lebensraum/11831242.html. Accessed 18 May 2018
4. Electronics practice (2016) Will the networked car be a living space of the future? http://www.elektronikpraxis.vogel.de/wird-das-vernetzte-auto-ein-lebensraum-der-zukunft-a-522714/. Accessed 18 May 2018
5. Bosch (2017) The car is becoming a new area of life. https://www.bosch.com/de/explore-and-experience/das-auto-wird-zum-neuen-lebensraum/. Accessed 18 May 2018

En Miniature: Our Living Spaces of the Future

13

Nowhere and Yet Everywhere

DoktorB Barbara Flügge

Abstract

Illustrations and design studies on the subject of mobile living spaces are a source of inexhaustible creativity. Strikingly, it is easy to imagine a mobile living space in the luxury segment. If it's not about luxury or comfort, but about creating affordable living space, what does a mobile living space look like for those who don't have a static place to stay? Those who do not have a home? Infrastructure designers, space providers, real estate operators, and mobility providers are amongst the key stakeholders that offer space. What qualities of entrepreneurial strength are needed to align a future scenario with moral and ethical norms and to draw the best possible scenario for the benefit of all?

The Comfortable Mobile Living Space

Illustrations and design studies on the subject of mobile living space are a source of inexhaustible creativity. Strikingly, it is easy to imagine a mobile living space in the luxury segment. Let's picture the following scenario. With the right equipment and technology, every module that docks on a vehicle's trailer offers an almost unlimited possibility of exchange (swap). Today, with the module living space on the way, I will travel in the morning and evening to and from my analogue workplace. If my employer offers a digital workplace, so much the better. A combination of modules could then be used, which are partly equipped as my living room and partly as my workplace. If this scenario is further

D. B. Flügge (✉)
digital value creators (DVC), Bottighofen, Switzerland
e-mail: barbara.fluegge@dvcconsult.com

B. Flügge (ed.), *Smart Mobility in Practice: Good Car, Bad Car, No Car – Is this the End of Nurturing Our Mobility DNA?*,
https://doi.org/10.1007/978-3-658-43171-6_13

developed, new forms of mobile home recreation areas emerge. In order to get from A to B and C, modules are given their own mobility track—similar to train tracks or commuting tracks—on the freeway. By induction technology, mobile charging stations provide the energy supply, and manage the loading and unloading cargo modules. Even the supply of food and consumer goods happens on the go.

Resourceful interior designers will create a living and working experience in the smallest of spaces, which, thanks to augmented reality and miniature home and office appliances, will prompt us to try this out for our next holiday. The management offsite then becomes an immediate inventory test in minimal space—the modules are lined up and ensure the virtual and digital networking of the participants. In particular, financially strong companies and event customers receive a premium service. This service includes special catering, coaching, and convenience services. The location of the mobile experience room is freely selectable. We are willing to pay for this luxury: whether this is a location with a fantastic view of a lake, or the peak of a Swiss mountain, like the Säntis, or having a nature and animal experience "within reach". We are happy to invest in and pay for special experiences.

If we are missing something, we can correct this by building up a stock in the local 3D printing terminal or in a service terminal. Our drone concierge can then deliver it directly to our kitchen or living room. We don't worry about the function of the bathroom or the bedroom module. The providers of caravanning are already experienced in these topics and already offer solutions today. We also no longer have to worry about traffic jams and the possibility of being late to work. The digital assistant ensures that every moving module stops at a proper time. And that these stops, and any other movement, are integrated smoothly into the traffic flow. The digital assistant takes care of all other private and professional matters.

This includes the selection of the nearest or optimal work order, networking with business partners, who should be met at the next but one stop, healthcare and the on-boarding of a doctor, if a digital health check now requires an on-site meeting. For those who do not know where they are allowed to stay, the digital companion takes care of all the paperwork required for the stay. He compiles the relevant information, interacts with the digital support centre of the relevant authority and gives us feedback accordingly. We don't have to worry about finding a parking space: there are private places beyond the access roads. We work at our mobile workstation, are perhaps on our way to a meeting, and, in the evening, we move to our next rest stop.

We can either buy or lease new modules. The personal assistant does not only spontaneously offer corresponding providers that match our desires with offerings. The assistant shows us virtually how the module can make our everyday life even more tangible. Augmented Reality feeds our life directly into the camera and ensures a direct recording of our empathic values by means of sensors, transmitters, or neuronal screening. The real-time triggering of the order speaks for itself. If we allow future planning in our past, the module will be on site tomorrow before we order it today.

Another aspect of the mobile living space lies in our almost limitless longing for the new. In today's digital age, individual consumer groups are increasingly looking for confirmation, experiences, and moments of surprise. The car as a moving or standing medium is just right for us here. The experience concierge might serve bundled offerings that are matched to the respective target group. In addition to brand, interior and module choices, our hungry heart pounds for driving autonomously, keeping up with the marketing promise of waking up where it's best for us. It's the surprise that counts! Where? We don't know yet. And that further raises our expectations. And excites us even more.

I have a call to action for you, dear decision takers. I would like to submit to you a proposal for the creation of mobile moments of enthusiasm! I had applied for an internship at a well-known interior design studio in eastern Switzerland and asked for a 3 month project. My application was probably dismissed as a spinning mill—I had received no more feedback: Why should an experienced and successful professional in the field of networking and digitisation (me!) be interested in an internship with an interior decorator?

Habitats Without Freedom of Choice

One of the projects we have covered in this book is ENTOURAGE. Initially, we assumed—in the context of this project—that the design of our living spaces in a mobility environment is left purely to us. With a few clicks, a wink of the eye or a thumb up, we signal what we want and our digital or virtual counterpart fulfils our wish. It is us who dictate, to a certain extent, which apps we store, keep, and remove from the device. In the future, I suspect that the transition from vehicle to vehicle, from medium A to medium B, will work independently of city and country borders, and that we will be giving seamless and device-independent "app service continuity". That's the proper way to provide seamless integration—you wouldn't disagree, right?

However, we should also admonish ourselves and ask critical questions regarding the use of our data. How much is giving others an insight into our behaviour and preferences worth? There is a commercial from Apple that turns the responsibility of data care over to the hands of consumers and device users. If there were a barometer which demonstrated the reuse of our data, and how it is monetised through this reuse, we would act differently! Even further, we would act fast.

Mobile living spaces find their buyers where there is budget available. But there has also always been, and always will be, groups of people who have to manage carefully with a tight budget. Would John Jakes' [1] outlook on the twenty-first century also stand the test of time? Would Jakes offer mobile living spaces to those groups of people who are forced to be on the road? Through John Jakes' treatise, we have inevitably projected the image of the streams of refugees into his narrative. Jakes' scenario should give us all, who grow up in prosperity, reason to think about those who are rejected by the modern world today, as well as about groups of people who simply can't afford certain places to live because of rising rents, living costs, and energy prices. If one were to apply John Jakes' scenario to today's discussion about the reception and settlement of refugees, could future designers find a solution here? Might there be a modern form of non-binding reception? Could permanent moving modules be established at certain crossroads of countries that, positively speaking, can provide a mobile and humane refuge?

Had John Jakes been able to make himself heard through social networks, discussions, and developments surrounding the mobile living space would have flared up at a much earlier point of time. A film producer would have sought the rights and set an exclamation mark with a top-class cast, presenting a deliberately critical debate concerning homelessness, emphasised by excluding a happy ending from the narrative. Perhaps we would then have put ourselves more directly into the position of vulnerable groups and revised opinions and actions on the basis of the vivid descriptions of John Jakes' novel's characters.

How are those who are unable to choose their preferred living space because of money, territory, or other unsolicited reasons coping? Imagine that autonomous living modules are preconfigured to an extent, where features, functions, expectations, and desires are subject to a predetermined construct. Budget-wise, statistically calculated needs are covered. Consumers would not long for further functions, such as the feeding of current music titles, an update of journalistic publications, different colours, or a mobile health check! The further development of such functions and applications is not necessary in a scenario characterised by satisfying the statistics and the calculated degree of saturation. In case of budget shortages, we can go further in our scenario: even distance, speed, and location swapping would not be possible. Unless the consumer pays for the service upgrade. The degree of saturation—how external stimulation steers us—seems programmed with foresight. Is that the way we want to treat each other? And do we really want to dictate what is being offered under what circumstance? Imagine that a system informs you of what is offered by whom, on which route and in which vehicles, and gives you the option of paying or leaving. It remains to be hoped that the human discoverer gene will stir and create diversity!

Reference

1. Flügge B (ed) (2017) Smart mobility – connecting everyone. Springer Vieweg, Wiesbaden

Moving on Together!

During the course of this book we have elaborated, scouted, and presented to you, in an exemplary and comprehensible way, how a responsible basic attitude coupled with an innovative and implementation-oriented application of existing and planned technologies affect our living spaces.

We have and will continue to have a lot of discussion about the roles of cars in society, politics, businesses, and private lives. Moreover, discussions will also consider technological progress, innovation and visions, misconduct, new, missing and required regulations, and driving bans.

We also need to rethink the impact of technologies on our own behavioural DNA—the car itself is now in our blood, heart, and we struggle to skip the car, sell it, or share it with others—this becomes impossible once we own the car and get used to it.

The flip side is that we have been nurtured by the automobile industry over decades, with luxurious and unnecessary features, with great inventions, allowing us to feel good and drive safely. Toddlers say "car" or "auto" first before saying "mama" or "papa". As children we recognise the car brand and the shape of the car. Mobility has become a genetic disposition, and the automobile and car industry nurture our "mobility-means-car" DNA in a very smart and creative manner.

It would be presumptuous to assume that traditional leaders in the vehicle business, industrialised countries and innovators with budgets alone should determine, or even end the course of the successful car trade. However, it is not the time of managers anymore, but instead of leaders. And the example of the aCar in Chap. 6 shows us how it can work.

In order to dress up the car and transform it into new spaces that go far beyond locomotion, we need a holistic, 360° approach. To face and tackle the challenges of the automotive sector downsizing hundreds of thousands of jobs, Smart Mobility is the means to invent, create, and generate 10 times more job opportunities in the field of digitisation, service creation, and persona-driven adventure routes. It is a question of assertiveness and therefore passion, our dedication and our willingness to take risks, and an urge to break common grounds to move on.

B. Flügge (ed.), *Smart Mobility in Practice: Good Car, Bad Car, No Car – Is this the End of Nurturing Our Mobility DNA?*,
https://doi.org/10.1007/978-3-658-43171-6

Should we abandon the car and step back in order to move on? Let's get back to the aCar: it is a great example, even a role model, of how human creativity, technological progress, and the car itself turn into a great starting point to "Move on". Let's be creative, stop worrying and take the aCar example to transform obstacles into chances—similar to the ladder that our students are climbing up on the ABC Business Opportunity Design course. I would like to thank my co-authors and interview partners, the Springer editorial team and all our readers who are taking up the opportunities and challenges of intelligent mobility!

The aCar

The acceptance of electromobility in countries like Ghana and Nigeria, for example, is different from that in industrialised countries: it is higher! The effort for mobility in countries like Ghana is calculated in number of hours—hours that the people in Ghana are able to walk: mobility is not based on horse powers, but hours and distance travelled on foot! A reach of the aCar of 80 km is therefore an enormous improvement compared to the effort required to cover this distance on foot! Almost too much to dream of when we are comparing it with the shorter distance from the village to the nearest market in the city. Isn't it the famous 80–100 km distance that worries us in Europe about taking an electric car and then being stuck in the countryside without a charging station? Back to Ghana: the energy storage system is the driving factor of the electric vehicle, the aCar. The aging of the battery correlates with the still available power of the battery. For the operation of a cooling unit or cooling compartment, a small energy requirement is sufficient to achieve a temperature of 0.5–1 °C. The temperature of the cooling unit or cooling compartment must be kept at a minimum. As things stand today, no products requiring refrigeration can be sold in the markets in Ghana. The technical devices are missing. This is where the hour strikes for regenerative energies and the use of the remaining energy storage of the electric vehicle. We are flipping this now. The vehicle aCar thus becomes an energy generator and accumulator! The electric vehicle is the optimal technical solution for Ghana and Nigeria and other markets for the aCar, for markets where there is no filling station network and where people are willing to pay for solar power.

Solar kiosks are now well established in developing countries. Solar energy is easily reproducible. A 10 % unused battery capacity of the aCar could lead to on-site processing and temperature-adapted storage of goods in the case of cassava farmers. The flour would be such a result of on-site processing. Instead of transporting the cassava plants to the interim storage facility and thus risking contamination during the journey, in the interim storage facility and in the production facilities, the flour is produced onsite. There are further advantages: the flour has a better quality. The farmer also has a higher yield ratio—the scrap that had to be expected previously due to transport and storage damage is eliminated.

Looking to the future, we envision further forms of suburban processing in rural areas of industrialised and developing countries: supported by Micro Energy Hubs and Micro Logistics Hubs. They promote inventiveness in the first stages of the value chain of a production chain and at the same time ensure new forms of local service provision.

Glossary

1-out-of-3 1-out-of-3 describes a mobility mix consisting of 3 means of transport. The offer of 1-out-of-3 requires the clarification of financing, legal requirements and the development of an adequate business and operating model. The effect formula is suitable for processing an adequate business model.

1-out-of-4 1-out-of-4 describes a mobility mix consisting of 4 means of transport. The offer of 1-out-of-4 requires the clarification of financing, legal requirements and the development of an adequate business and operating model. The effect formula is suitable for processing an adequate business model.

aCar aCar stands for "any car". aCar is a multifunctional electric vehicle and is the result of a research project launched by the Technical University of Munich. The newly founded EVUM Motors GmbH based in Garching is responsible for the market launch and series maturity.

Ambient Living The increasingly cell-like design of autonomous vehicle shapes allows the interior of a vehicle or cell to be adapted to the wishes and habits of passengers. The adaptation takes place in digital interaction with the Digital Concierge, the personal assistant of the mobility user. When booking the guest, the vehicle processes the corresponding preference profile in the concierge and uses the space in the car for entertainment and knowledge transfer, relaxation, or shopping.

Application/Use Case The term application or English *use case* is often used synonymously with usage scenarios. Originally introduced by Ivar Jacobson in the year 1992, the term Use Case was used to describe the requirements of a software system in software modelling using the description language Unified Modelling Language (UML).

BIM Smart Mobility consists of a multitude of strategic, functional, and technical elements. These elements are the so-called Building Blocks of Intelligent Mobility (BIM). The elements for project and change management are combined under the Smart Mobility Program Management. The strategic and project-related elements it contains support every project, regardless of its technical, functional, or process-related focus. We assign four levels to each technical and functional element: Smart Services, Smart

B. Flügge (ed.), *Smart Mobility in Practice: Good Car, Bad Car, No Car – Is this the End of Nurturing Our Mobility DNA?*,
https://doi.org/10.1007/978-3-658-43171-6

Data, Smart Products and Smart Spaces. The BIM is used in the Smart Mobility procedure model.

Context The context depicts the situation in which the organisation operates or wishes to operate in a given state: now, in the future, in a specific timeframe or geography. Herein, which organisations can take which position in a web-based business space, the service marketplace, is situation dependent or—in other words—contextual.

Convergence in relation to mobility With respect to mobility, we are talking about the convergence of passenger and freight transport. Therefore, traffic flows, the utilisation of space, assets, and services that are key to passenger transport and key to freight transport are being connected. Unlike isolated systems and asset management, convergent mobility helps to reduce noise, air, and light pollution.

Concept car A concept car translates a company's claim into a haptic, tangible structure in the form of a mobile, operational vehicle. The vehicle is usually built once. Ideas, innovative features and functions, as well as the progress in development can be showcased, functionally tested, and presented in the concept car. Those become visible. The concept car serves as a tangible yet living object for the company's internal divisions, business partners, and consumers.

Customer Journey The Customer Journey describes a typical daily routine of a previously agreed person or persona. The description is made via a precise schedule, the activities recorded in it, the associated emotions and a recording of the touchpoints used. Touchpoints are the media we use to collect or retrieve data. Examples are smartphones, wearables, and web portals.

Decision Triangle The Decision Triangle of Smart Mobility describes the personally perceived optimum of comfort, time and budget that determines the focus on mobility.

Design Thinking Established methods and procedures for the analysis and definition of interactions have come to light through the introduction of design thinking as a design element for product and idea development.

Digital Concierge The Digital Concierge is the personal assistant or digital companion. He serves and controls us during travel, selection, booking, use, evaluation, and deselection of services. He supports us end-to-end or door-to-door in our daily and activity processes. Digital companions create experience spaces in the car and outside. In a vehicle, the concierge already processes the corresponding preference profile at the time of booking and pre-entering the vehicle. The concierge serves the passengers with entertainment and knowledge transfer, relaxation, or shopping services.

Digital Consumer The Digital Consumer is a Persona that we encounter in digitisation projects. The archetype of a Digital Consumer initially appeared in projects of Industry 4.0 and the Internet of Things (IoT). Similar to a human being, the Digital Consumer orders, consumes and offers data as well as status information. Examples of Digital Consumers are sensors, wearables, intelligent fabrics, supplier parts of machine-made products and/or the product itself.

Digital Economy A definition widely used in German-speaking countries is that of the synonym "internet economy": "Internet economy is a predominantly digitally based economy that uses computer-based networking to enable communication, interaction and transaction in a global environment" (Lies (2013). Gabler Business Encyclopedia. Keywords: internal communication, online on the Internet, Springer Gabler, URL: http://wirtschaftslexikon.gabler.de/Archiv/326738/interne-kommunikation-v3.html, accessed 08.01.2013, last visit: 19. May 2018).

Digital Ecosystem A Digital Ecosystem is a functional habitat that consists of individual and organisational entities, organisms, infrastructure, assets and all non-living natural and anthropological environmental factors that all interact with each other and with the environment in energetic, value and service-oriented, textual, informative, and digital cycles.

Digital Diversity In servitisation, the demand for personalisation is increasing to the same extent as the exploding development of individually produced or personalised consumer goods. See also Exponential Digital Diversity.

Ecosystem Analysis See Ecosystem Assessment

Ecosystems Assessment The Ecosystem Assessment is an exploration of ecosystems and their local, known, and unknown, virtual, and physical market participants, customers, and business partners. Examples of Ecosystems are sites, enterprises, corporate subsidiaries and headquarters, urban and rural infrastructures. Even a street, a field trial, a digital co-working space, and an e-sports event are great growth and mobility areas of investigation and applicable for Ecosystem Assessments. Stakeholders and members of the ecosystem follow a four-step workshop that runs over one to two days.

An Ecosystem Assessment involves the investigation of the origin, objectives, status quo and growth capacity of an ecosystem. This includes the analysis of known and unknown stakeholders and the identification of essential persons. This takes place on the basis of the composite analysis. In addition to service design and development, the aim is to test the service trial in a suitable room.

Ecosystem Initiation—Driving Forces of Initiating an Ecosystem Ecosystems are driven by a minimum of one of the following forces and intentions:

- Social bonding: having the mandate or urge to preserve the societal belonging of individual, institutional, and organisational entities.
- Value driven bonding: maintaining a community of values prior to, during, and after an incident or event.
- Commercial bonding: service and/or product supply as well as service and/or product deployment imply further ecosystem relevant actions.
- Purpose driven bonding: being an economic, political, or timing-related purpose that leads to a multi-sided collaboration in an ecosystem.
- Incident driven bonding: an incident raises the sense of urgency and asks for a resolution.

Ecosystem Resourcing Ecosystem Resourcing denotes the process of equipping the habitat with life essential, operations relevant and functional resources. Herein the process of turning the habitat into a self-contained functioning system is factored in. In case of a lack of supply or bottlenecks arising due to emergency measures and natural disasters, the risk for the individual increases.

Effect Formula The Effect Formula determines mobility costs, implementation costs, and quantifiable benefits of mobility projects. We introduced the Effect Formula for the first time in the MaaS project.

Experience Feeling and Experience-Guided Mobility In the future, experiential mobility will determine which means of transport and which degree of automation we prefer for certain routes. In the future, mobility that feels like an experience will determine whether we actually use the means of transport used again; the basis for this is the assessment of our perception, i.e. whether the use has left a positive or negative impression on us.

Exponential Digital Diversity In servitisation, the demand for personalisation is increasing to the same extent as the exploding development of individually produced or personalised consumer goods.

Green Class The Green Class is a project of the Swiss Federal Railways (SBB), OEM company BMW and PubliBike with the aim of offering a mobility mix of rail, electric vehicle, and bicycle services. After the first year of testing, it was decided in 2018 to continue the Green Class.

Growth Space With the term growth area, we are talking about locations and spaces that have growth as their goal based on existing structural characteristics and a clearly defined strategy. There are geographical and virtual locations and spaces.

Habitat In analogy to ecosystems, the quality of a habitat, regardless of its size, geography, and social demands, is determined by the degree of togetherness and reciprocity, access to resources and facilities, and the sensitive use and employment of the existing habitat.

In-Car Living In-Car Living annotates the car as a living space. The car is increasingly competing with other space sites such as offices, homes, shops, and social meeting places. Another term coins the relevance and the financial impact of cars. There is another extended viewpoint: denoting In-Car as "Wheel Estate" in analogy to "Real Estate".

In a socio-critical context, we are referring to In-Car Living as a refuge, a place for the homeless and people without a home community, people that cannot afford living costs and real estate. John Jakes' novel depicts this scenario in his dystopia "On Wheels".

In our chapters "Mobile Adventure Room" and "En Miniature", we expand the reference personas and the context through a new life experience in the car.

Mobility Analysis We encounter in the market different levels of details with respect to retrievable mobility data for smart analytics. Data about new user groups, influencers, and tactical stakeholders, mobility means like escooters and others, are not even made available nor accessible. We recommend that at least one mobility analysis is thoroughly carried out, including a demographic study and analysis of income streams by distinct parameters. Those serve to examine the dependencies between mobility consumption of public transport and the ownership of private cars. Furthermore, a persona-dependent route analysis will serve to identify the motifs for opting or deselecting a transport mean.

Mobility-as-a-Service (MaaS) MaaS customers pay for mobility per itinerary or monthly depending on mobility usage, distance, or other mobility criteria. The packaging of mobility needs results in a subscription-based model where the MaaS customer pays an amount of money per mobility package. The price for the package derives from the combination of transportation means and the number of hours that have been spent on the public bus and the mileage spent in the taxi or a shared or rented car.

Mobility Provider A mobility provider encompasses any entity that offers mobility—be it for passengers or cargo. These providers have the right to offer mobility as a consumable service—regardless of whether mobility is offered by private, public, or shared means of transport, and regardless of whether and which provider serves individual or entire routes. Possession of a means of transport is irrelevant.

New Role Models—The Mobility Manager The *mobility manager* sets the ground for all mobility relevant requirements and their fulfilment within an ecosystem and the relevant locations. Driven by the assembly of an ideal location profile, the mobility manager is able to target private and public service offerings in such a manner that mobility is made consumable and financeable. Based on our projects we composed a competency and skill training for becoming a Mobility Manager.

New Role Models—The Project Incubator The Project Incubator is an organisational unit that triggers innovation. It is the nucleus of creation and innovation. It creates the right creative atmosphere and brings the different people together, who then seriously deal with the projects and promote the growth of the germ cells with sensitivity. A further core competence lies in bringing together founders and mentors, idea finders and idea promoters. The project incubator deliberately works outside-of-the-box. They are a forward thinker who succeeds in enriching prevailing mechanisms of collaboration on site with a wealth of ideas from individuals, initiatives, companies, and research. The performance of the role is then successful if the results of innovation management can be presented, understood, and implemented.

Similar to the Ford Smart Mobility program, the project incubator has an overview of all activities. Only then can multiple efforts be avoided or the rejection of ideas that were not successful at other locations be accelerated. Conversely, a network of innovations can be established by comparing them with other ecosystems. This allows the effort and acquisition of necessary skills and competencies to be shared.

Another core competence of the project incubator is the acquisition of, and cooperation with, start-ups, entrepreneurs, and innovation departments of companies.

New Role Models—The Self-Sustaining Ecosystem The Self-Sustaining Ecosystem withstands adversities such as risks, external interventions, and unforeseen events. It is anchored in itself and organises itself as independently as possible. A model that examines the acceptability of organisations in the context of an ecosystem was introduced in the book *Smart Mobility – Connecting Everyone*.

OCN User Groups The acronym OCN stands for Occasional, Coincidental, Nomads. These are user groups, i.e. personas, who are becoming a target for service providers through emerging mobility concepts and offerings. Occasional stands for those that do not commute or drive/co-drive regularly on one route. Coincidental depicts random passengers who get excited by an experience and therefore kick-off a mobility trip. Nomads do not have a fixed living place or habitat. They travel quite intensively from one place to another. Digital nomads travel extensively while working remotely through technical and digital means.

Omnipresent Mobility Omnipresent mobility means that we could be at any time in the world. Being omnipresent means being able to move and decide freely. Information technology and the art of engineering contributed substantially to the evolution of omnipresent mobility.

On-off-Adventures It is our choice which services we would like to consume or not. The On-off-button is not necessarily a manual button that needs to be pushed physically, but it could be steered by audio, eye contact, our past experience analysis, and our taste. Thanks to digitisation and systems-controlled preference control, experiences are comprehensible and can be offered within seconds to promote decision-making. Our own and environmentally relevant moods draw a picture of our taste and translates them into digital services and where these services are then deployed.

Persona In the sense of design thinking, a persona describes an archetype, which is composed of similar or consistent behavioural patterns of consumers. Examples of archetypes are the persona "consumer generation from 20 to 25 years of age", the persona "recurring cruise passenger", the persona "analyst at the transport company", and the persona "business traveller". The archetypes offer an image of comparable consumers or user behaviour after detailed research of motivation, wishes, expectations, and needs. In our tools, we expanded the Persona principle to further human and non-human beings, structural and virtual items.

Persona Profiling See Persona Profiling Setcard

Persona Profiling Setcard The persona's profile and preferences are registered in your Persona Profiling Setcard. We created the card game to serve as a basis for any kind of persona/archetype enhanced research, field study, solution design, and brand refinement. Getting to know your user and consumer—regardless of whether this is a human or an avatar, a thing or a data collector—is indispensable in digital transformation and digitisation projects. Essential functions and features are detected. These are valuable insights in business process design and are particularly valuable when they enter the business process designs at the right time. The Setcards serve alongside the scrum phases for the development team.

Personas or Personae Plural of Persona.

Segmentation A segmentation of mobility offerings, e.g. into public and private mobility offers or target group specific assignments, such as for customers and business partners, offer constraints for perspectives. Particularly in times of rising infrastructure burdens and increasing competition between regions and cities, an isolated view is no longer appropriate nor helpful to resolve the challenges in a hyperconnected world.

Companies and administrations are interested in the influence they can have on mobility in the market. The strategic positioning of the company for a city or region comes to the fore. The approach of the service-oriented trade under the application of the Internet of Services supports the coverage of the whole complex. Certainly, techniques and methods are required to make this complexity comprehensible and step-by-step implementable. Our experience has shown that this is possible.

Self-Sustaining Ecosystem The starting point for service-oriented action is the claim of market participants and consumers to keep their living space and economic structure alive in the best possible way.

Service The term Service by nature can be technical, business related, or craft related. In German we distinguish technical services (Dienst) and business services (Dienstleistung). In other languages you do not always find a clear distinction.

For the term Service there are different ways of looking at it. Consumers have their own opinion when it comes to the concept of service and equate it with the concept or offer and/or provision of a service. Services are regarded as an asset, i.e. as a gain, whether in the sense of a planned, requested service or a service that supplements a product delivery as an accessory. Irrespective of this, the consumer demands a consistent Service Orientation of the supplier from the idea up to the implementation and consumption.

Service Aggregator Aggregators such as the Swiss Federal Railways with the General Abonnement offer unlimited use of buses, trains, ships, and public facilities for an annual fee. Innovators such as the Mobility-as-a-Service initiative offer profile-based, i.e. persona-dependent, service bundles. The latter define a bundle of rental cars, buses, and trains depending on the useful life of a public transport system and the target number of kilometres to be driven. Pricing is thus based on utility-related business models.

Service Provider The service provider is in charge of projecting consumer demand in an utmost and hopefully creative manner. The more efficiently and effectively the matchmaking process is safeguarded by information technology, the higher the probability that the original demand of the consumer is fulfilled by the product or service that is being delivered.

Service Broker Mobility brokers range from transportation and passenger transport companies to hospitality providers such as hotels, restaurants, tourist offices, travel agencies and shopping malls, VIP agencies, carsharing, and leasing providers. Digital brokerage is subject to platform providers and marketplace operators. Those that operate comparison portals and search engines broker search and pricing services.

Service Channel Maker Consumption channels for mobility will be more diverse than ever: whether this is an app, a multifunctional check-in terminal at home, or a smart bus stop. Channel Makers are often perceived as being the real innovation drivers, working on behalf of established institutions or as start-ups, to create new docking points for customer acquisition and retention. The higher the reach and number of users, the higher the conversion rate, i.e. the actual booking of the offer.

Service Consumer Mobility Consumers are diverse. Mobility Consumers range from individuals to organisations, and to groups such as travel groups and families. On the organisational side, companies and public administrations use mobility services. Other consumers evolve through the Smart Mobility Ecosystem and its service expansion technique.

Service Gateway Service providers and hosting rely on Service Gateways, both as in-house or third-party providers such as IT companies and system integrators. They ensure the interoperability of data, processes, and services. Data-as-a-Service, not only ensures interoperability, but also helps the gateway to an expanded business model.

Service Hosting Virtual services, such as ticketing, travel and transportation of goods, and packaging are hosted by service providers and/or third-party providers on the web or via apps, at the virtual counter and in the transport means themselves. Publicly available physical services are hosted by railways, roads, or bridges.

Service Marketplace The service marketplace, often mistakenly referred to as a shop or store, is a web-based business space. Selection, ordering, purchasing, collaboration, strategic alliances and market-specific analyses and offers take place in this web-based location, i.e. they are digitised. Devices such as smartphones, tablets and wearables offer access to the marketplace. The order is placed via digital and physical consumers.

Service Orientation Mobility will decisively develop into a service-oriented structure of effects, in which roles and responsibilities, as well as other functions, change. Based on a role model in the sense of the Internet of Services, they offer space for the preparation and reorientation of previous strategic, functional and task-related positions.

Servitisation The starting point for service-oriented action is the claim of market participants and consumers to keep their living space or economic structure alive in the best possible way. The approach to the optimal service offer for the respective context goes hand in hand with the degree of tradability of services. This is where physical or digital media and the underlying interactions (links) of a service take effect. The service becomes therefore both a motivator and the result of interaction-related analysis.

Smart Mobility Ecosystem By encompassing the entire Smart Mobility Ecosystem rather than prioritising one facet or platform, it helps us see the bigger picture: seeing the interconnection between the digital and the physical worlds, addressing both people and cargo needs and understanding the underlying Building Blocks for Intelligent Mobility (BIM) as well as the new infrastructure all provide this insight.

Smart Mobility Mobility is then smart as soon as projects are implemented according to the following criteria: (1) beneficial for all concerned, (2) manageable for public and private infrastructures and (3) for the benefit of the ecosystem. It is not without reason that the successful implementation of intelligent mobility concepts by those responsible for Smart City, Smart Nation, and regional projects (Smart Regions) is one of the most important critical success factors.

Smart Mobility Procedure Model The successful implementation of Smart Mobility applications stands and falls with a structured and yet creative approach. The Smart Mobility procedure model captures the different motives for dealing with Smart Mobility. A fundamental distinction can be made between six project planning approaches:

1. Explore new business areas, review existing business models and/or explore one's own position in the context of mobility.
2. Review, enhance, or expand existing building blocks of intelligent mobility with regard to possible market offerings.
3. Initiate and implement the Smart Mobility innovation project once the decision has been made.
4. Carry out a fitness check for your own ecosystem in the sense of "How fit is my city, district, or venue in terms of Smart Mobility?"
5. Identify positioning opportunities for own technologies and/or ideas and market access opportunities.
6. Conduct a context check to compare your own ecosystem with one or more ecosystems. This can be applied independently of Smart Mobility.

Stakeholder and Stakeholder Analysis A stakeholder analysis provides insight into members and participants of an ecosystem. This is independent of the size or geographical extent of an ecosystem. Experience shows that, in particular, the analysis of the participants by external parties is aborted too early. Even if 20 or 100 participants are encountered, this might be too early. It is precisely the increasing extension to further personae that reveals mission critical information about obstacles, chances of success and further service offers.

Survival Space In our context, the survival space describes the car as a critical resource that ensures the survivability of both owners and users. This works through the multifunctionality built into or offered by the vehicle: it enables transport, storage, cooling, and warehousing of raw and half-finished products, power supply, and additionally safe travel. Survival spaces become an essential part of the value chain for raw material processing—taking for example into account the European Supply Chain Act.

Touchpoints Touchpoints are the access points and media that we use. Usability designers use those in the process of collecting and retrieving data, executing functions on diverse devices such as smartphones, wearables, and web portals.

Track Ownership Analogous to airspace control, the route ownership determines the route control for urban areas. This will involve regulators, space, and route allocations as well as billing models. There, the monitoring and securing of ground and air space takes place.

Usage Scenario See Use Case.

Use Case The usability to translate test beds and pilots takes place in usage scenarios. Use case is a synonym for the term usage scenario.

Index

B. Flügge (ed.), *Smart Mobility in Practice: Good Car, Bad Car, No Car – Is this the End of Nurturing Our Mobility DNA?*,
https://doi.org/10.1007/978-3-658-43171-6

SPRINGER NATURE

GPSR Compliance

The European Union's (EU) General Product Safety Regulation (GPSR) is a set of rules that requires consumer products to be safe and our obligations to ensure this.

If you have any concerns about our products, you can contact us on ProductSafety@springernature.com

In case Publisher is established outside the EU, the EU authorized representative is:

Springer Nature Customer Service Center GmbH
Europaplatz 3
69115 Heidelberg, Germany

The manufacturer's authorised representative in the EU is Springer Nature Customer Service Centre GmbH, Europaplatz 3, 69115 Heidelberg, Germany. If you have any concerns regarding our products, please contact ProductSafety@springernature.com

Printed and bound by CPI Group (UK) Ltd, Croydon, CR0 4YY
01/10/2026
02225893-0020